Digital Circuits and Devices

Roger M. Kersey
Senior Instructor, National Education Center
Livonia Campus

Jointly Distributed By:

A Subsidiary of National Education Corporation

John Wiley and Sons, Inc.
Publishers
605 3rd Avenue
N.Y., N.Y. 10158

Library of Congress Cataloging in Publication Data

Roger M. Kersey
 Digital Circuits and Devices
 Electronics Textbook Series 4860
 Includes Index.

 1. Digital Circuits   IV. Title.   I. Series
Library of Congress Catalog Card Number: 84-62507
ISBN 0-93-318704-1

Digital Circuits and Devices
Copyright © 1985 by National Education Corporation.

All Rights Reserved. No part of this publication may be reproduced in any form, electronic, or mechanical, including photocopy, recording, or any information storage and retrieval system, without permission in writing from the publisher.

Vice President, Product Development: JoAnne Schiller
Managing Editor: Gary K. Edwards, Sr.
Editors: Arvey I. Andrews
         Gerald Kramer
Production Coordinator: Frank Wassil
Cover Art: Dave Korobkin
Illustrations: Charles T. Tokash

ISBN 0-93-318704-1

PREFACE

This book will provide the reader with the concepts of digital electronics which can be used as the basis for the understanding of computer and control systems. The topics are arranged in a very logical proven order allowing the reader to develop an indepth understanding of digital electronics in the shortest amount of time.

The reason for the development of this textbook was to bridge the gap between technical concepts and engineering concepts used in digital electronics and computer systems today. The text was written in a pyramid format allowing the reader to achieve any level in digital electronics that is desired. All chapters start off at a very low technical level, and as the reader becomes more familiar with the material, the technical level becomes more advanced to the point of reaching a beginning engineering level.

This text was written to be used by two and four year colleges and technical schools that wish to provide a comprehensive study of digital electronics. The book can also be used by anyone who wishes to update their skills in electronics.

The prerequisites for this text are a basic knowledge of the following subjects: AC, DC, transistor concepts, and the use of basic electronic test equipment.

With the data sheets, and general information found here, this text can also be used as a reference book as well as a standard text book.

INTRODUCTION

Digital electronics as we know it today, is based on the concepts of logical deduction that was first developed by the Greek philosopher Aristotle. The word digital refers to a type of process that uses discrete units. The discrete units may be true/false statements or *high/low* and on/off levels. Therefore, this form of electronics involves performing a logic decision or task based on its discrete inputs.

Around 1850, a self-educated English mathematician named George Boole applied Aristotle's deductive logic to mathematics. Boole applied mathematical concepts to logic principles and came up with a logical/mathematical system known as Boolean Algebra. Boolean Algebra is used in digital electronics to explain the operation of digital circuits and to reduce the size of a circuit while retaining the original function of the circuit.

The digital concept of Boolean Algebra was first used in industrial controls in the late 1800's to control machinery. The digital logic used relays to determine what type of logic function would be processed.

In the early to mid 1900's when receiving tubes were developed, the logic devices were switched from relays to receiving tubes. The reason for the change was that tubes were able to switch states faster than relays. At this point in time, the term digital electronics came into being.

The next major step in the evolution of digital electronics occurred in the late fifties and early sixties when the first semiconductor device was developed. This has led to the popularity of digital electronics today. With the advent of semiconductor technology, the number of functions an IC could contain increased from one type of function to over 100,000 functions in the same area. As the amount of integration increased, the cost, size, and weight of digital circuits decreased while the speed and flexibility increased. As a result, it is important for every technical to have a working knowledge of digital electronics.

Each chapter in this book covers various digital topics. These topics are then divided into distinct subtopics where needed, and covered in greater detail. The chapters begin with a list of objectives which describe what the reader should be able to do upon completion of the chapter. There are also topic review questions throughout providing the reader with a quiz of the material. At the end of each chapter is a list of summary points stating the important concepts followed by a chapter progress evaluation.

Chapter 1, **Digital Concepts**, describes basic digital terminology, analog and digital signals, and the advantages and disadvantages of both types of signals.

Chapter 2, **Numbering Systems**, introduces the reader to the many different types of number systems used in digital electronics, computer data coding systems, and digital/computer math.

Chapter 3, **Digital Gates**, introduces the concepts of basic logic functions and the development of truth tables, timing diagrams, and Boolean expressions for the basic logic gates.

Chapter 4, **Logic Development**, allows the reader to develop the ability to derive complex truth tables, gate circuits, and Boolean expressions.

Chapter 5, **Logic Families**, describes in detail the operational concepts of semiconductor logic as well as the advantages and disadvantages of each logic family. Also included in this chapter are logic family terminology, how to read and use specification data sheets, interfacing between logic families, and positive and negative logic operation.

Chapter 6, **Logic Simplification**, allows the reader to develop the use of Boolean algebra and Karnaugh mapping to reduce the size of the digital circuit while retaining the circuit's original function.

Chapter 7, **Digital Integration**, describes the operation of complex multi-gate devices that perform special tasks which include multiplexers, demultiplexers, encoders, decoders, displays, display drivers, and other complex digital devices.

Chapter 8, **Latches/Flip-Flops/Shift Registers**, introduces the concepts of digital storage and data bit manipulation devices.

Chapter 9, **Counters**, introduces the operational concepts and design of various types of digital counters.

Chapter 10, **Digital and Analog Conversion**, introduces the methods of converting analog data into digital data, as well as the conversion of digital data into analog data. The concepts introduced in this chapter will be the basis for computer and digital interfacing with the outside world.

Chapter 11, **Memory**, introduces the operational concepts of all types of digital storage, as well as the advantages and disadvantages of each type of memory unit.

Chapter 12, **Introduction to Computers**, introduces the basic internal operational concepts of computers and computer systems.

I would like to thank the staff at NEC Livonia campus for the helpful suggestions given to me for this book.

Thanks also goes out to all the manufacturers that allowed me to use data and specification sheets in this text.

Special thanks goes to Richard E. DeNagel, V.P. Product Development, and the staff at NEC Technical Publishing Division for their help, editing, and encouragement in the preparation of this text. I would also like to thank ICS/INTEXT for the data, pictures and diagrams used in this text.

Most of all, I would like to thank my wife, Regina and my son, Christopher for their extreme understanding and patience with me during the preparation of this book.

Contents

Preface iii

Introduction iv

Chapter 1 Digital Concepts

1-1 Introduction 1
1-2 Analog Signals 1
1-3 Analog Systems 3
1-4 Digital Signals 4
1-5 Digital Systems 4
1-6 Digital Terminology 5
1-7 Boolean Algebra 7
1-8 Digital Vs. Analog 8
1-9 Analog and Digital Systems: 8
     Advantages and Disadvantages
1-10 Summary Points 10
1-11 Chapter Progress Evaluation 10

Chapter 2 Numbering Systems

2-1 Introduction 11
     Counting in Binary
     Digit Position Weights For Decimal and Binary
     Binary to Decimal Conversion
     Decimal to Binary Conversion
2-2 Hexadecimal Numbering System 17
     Hexadecimal System
     Counting in Hex
     Hex/Binary Conversion
     Hex to Decimal Conversion
     Decimal to Hex Conversion
2-3 Octal Numbering System 22
     Counting in Octal
     Octal/Binary Conversion
     Octal/Hex Conversion
     Octal to Decimal Conversion
     Decimal to Octal Conversion
     Split-Octal Notation

2-4	Binary-Coded Decimal Numbers	26
	BCD Codes	
	The 8421 Code	
	The Excess-3 Code	
2-5	Coding Systems	29
	Gray Code	
	ASCII	
	EBCDIC	
2-6	Summary Points	31
2-7	Chapter Progress Evaluation	32

Chapter 3 Digital Gates

3-1	Introduction	33
3-2	The AND Gate	34
	Timing Diagram	
	Increasing Inputs	
3-3	The OR Gate	42
	Timing Diagram	
	Increasing Inputs	
3-4	Inverter	49
	Timing Diagram	
3-5	The NAND Gate	51
	Timing Diagram	
	Increasing Inputs	
	NAND Inverter	
3-6	The NOR Gate	56
	Timing Diagram	
	Increasing Inputs	
	NOR Inverter	
3-7	Exclusive-OR Gate	61
	Discrete ex-OR Circuit	
	Timing Diagram	
3-8	The Exclusive-NOR Gate	64
	Discrete ex-NOR Circuit	
	Timing Diagram	
3-9	Gate Inversion	66
	Sign Inversion	
	Sense Inversion	
	Sign & Sense Inversion	
3-10	Summary Points	72

3-11 Chapter Progress Evaluation   72

Chapter 4 Logic Development

4-1 Introduction   75
     Rules
     Boolean Expression Review
        Application of Rules
4-2 Deriving Gate Circuits from Boolean Expressions   80
     Rules
        Application of Rules
4-3 Deriving Complex Truth Tables   83
     Truth Table Review
     Application of Truth Table Principles
4-4 Deriving Gate Circuits from Truth Tables   86
     High Output Method
     Low Output Method
4-5 Summary Points   89
4-6 Chapter Progress Evaluation   89

Chapter 5 Logic Families

5-1 Introduction   91
5-2 Parameters of Logic Families   92
     Voltages, Currents, and Logic Levels
     Speed of Device Operation
     Noise Immunity
     Operating Temperatures
     Loading, Fan-in, and Fan-Out
5-3 Specification Sheets for Logic Families   98
     Manufacturers and Specs
     Cross Identification for Logic ICs
     What Specs are Important
     Pinout Diagrams
5-4 Saturated Bipolar Logic Families   102
     Resistor-Transistor Logic (RTL)
     Diode-Transistor Logic (DTL)
     High-Threshold Logic (HTL)
     Transistor-Transistor Logic (TTL)
5-5 TTL Subfamilies   105
     Standard TTL

	High-Speed TTL	
	Low-Power TTL	
	Schottky TTL	
	Low-Power Schottky TTL	
	Summary of the TTL Subfamilies	
	Open Collector TTL	
	Tri-State TTL Systems	
5-6	Nonsaturated Bipolar Logic Families	114
	Current-Mode Logic (CML)	
	Emitter-Coupled Logic (ECL)	
	Integrated-Injection Logic (ILL or I^2L)	
5-7	MOS Logic Families	117
	NMOS and PMOS	
	Low Threshold PMOS	
	VMOS	
	DMOS	
	HMOS	
	CMOS	
	CMOS Subfamilies	
5-8	Summary of Logic Families	120
	Interfacing Logic Families	
	Using Logic Families	
	Using TTL	
	Using MOS	
5-9	Negative Logic	122
	Negative-Logic Gates	
	Positive-Logic Through Negative Gates	
	Negative-Logic Truth Tables	
	Equivalent Logic in Diagrams	
5-10	Summary Points	126
5-11	Chapter Progress Evaluation	127

Chapter 6 Digital Simplification

6-1	Introduction	129
6-2	Boolean AND Identities	130
6-3	Boolean OR Identities	132
6-4	Boolean NOT Identities	134
6-5	Boolean Laws	136
	Associative Law	
	Commutative Law	

        Associative and Commutative Laws
        Distributive Law
        DeMorgan's Law
6-6   Boolean Theorems   139
        Theorem 1 (Common Variable Reduction)
        Theorem 2 (Complement Variable Reduction)
6-7   Karnaugh Mapping   143
        What a Karnaugh Map Can Show
        Truth Table Equivalents
        Minimizing Logic Circuitry
        Drawing Larger Karnaugh Maps
        Plotting Multiple Variables
        Looping Minterms Together
        "Rolling" a Karnaugh Map
        Reading Karnaugh Loops
        From Equation to Logic Circuitry
6-8   Summary Points   153
6-9   Chapter Progress Evaluation   154

Chapter 7   Digital Integration

7-1   Introduction   155
        Arithmetic Operations
            Binary Addition and Subtraction
            Parallel and Serial Arithmetic
            Basic Half Adders
            Full Adders
            Logic Subtractors
7-2   Multiplexers   161
7-3   Demultiplexers   164
7-4   Fundamentals of Encoding and Decoding   166
        1-of-8 Decoders
        Encoders
7-5   Digital Displays   171
        Light Generating Display Sources
            Gas Discharge Tubes
            Light Emitting Diodes
        Non-light Generating Display Sources
            Liquid Crystal Display
        Display Formats
            Multiple 7-Segment Display

	Dot Matrix Display	
	Multiple Dot Matrix Display	
7-6	Display Decoder/Drivers	178
	Single 7-Segment Decoder/Driver	
	Multiplexed Decoder/Driver	
	Dot Matrix Decoder/Driver	
	Multiple Dot Matrix Decoder/Driver	
7-7	Other Digital Integration Functions	184
	Expandable Gate Units	
	Schmitt Trigger Inputs	
	Digital Bi-Lateral Switches	
	Magnitude Comparators	
	Parity Generator/Checkers	
7-8	Summary Points	189
7-9	Chapter Progress Evaluation	190

Chapter 8 Latches/Flip-Flops/Shift Registers

8-1	Introduction	191
8-2	RS Latches	191
	RS NOR Latch	
	Hold Mode	
	Set Mode	
	Reset Mode	
	Disallowed Mode	
	RS Nand Latch	
	Disallowed Mode	
	Reset Mode	
	Set Mode	
	Hold Mode	
8-3	Clocked RS Latches	199
	Clocked RS NOR Latch	
	Disable Mode	
	Hold Mode	
	Set Mode	
	Reset Mode	
	Disallowed Mode	
	Clocked RS Nand Latch	
	Disabled Mode	
	Hold Mode	
	Set Mode	

	Reset Mode	
	Disallowed Mode	
8-4	Data Latches	204
	Disable Mode	
	Reset Mode	
	Set Mode	
	7477 Quad Data Latch	
8-5	RS Master Slave Flip-Flops	206
	Disable Mode	
	Hold Mode	
	Set Mode	
	Reset Mode	
	Disallowed Mode	
8-6	JK Master Slave Flip-Flop	210
	Disable Mode	
	Hold Mode	
	Set Mode	
	Reset Mode	
	Toggle Mode	
8-7	Data Flip-Flops	214
8-8	Shift Registers	215
	Types of Inputs	
	Types of Outputs	
	Types of Shifting	
	4-bit Sispo Shift Right Shift Register	
	Serial Input Operation	
	Outputting Data	
	Parallel Data	
	Serial Data	
	4-Bit Spispo Shift Right Shift Register (7495)	
	Write/Recirculate Shift Register	
	Shift Register Counters	
	Ring Counter	
	Johnson Counter	
	Additional Uses for Shift Registers	
8-9	Summary Points	225
8-10	Chapter Progress Evaluation	226

Chapter 9 Counters

9-1	Introduction	227

xiii

9-2	Standard Asynchronous Counters	227
	MOD 2 Counters	
	Asynchronous MOD 4 Counters	
	Up Counter	
	Down Counter	
	Up/Down Counter	
	Larger Asynchronous Counters	
	MOD 8 Up, Down, and Up/Down Counters	
	TTL 7493 Binary 4-Bit Ripple Counter	
9-3	Standard Synchronous Counters	237
	Synchronous MOD 4 Counters	
	Up Counter	
	Down Counter	
	Up/Down Counter	
	Larger Synchronous Counters	
	MOD 8 Up Counter	
	MOD 8 Down Counter	
	MOD 8 Up/Down Counter	
	MOD 16 Up, Down, and Up/Down Counters	
	TTL 74193 4-Bit Up/Down Counter	
9-4	Non-Standard Synchronous Counters	250
	Forced Reset Method	
	Cause/Stop Method	
	Design Procedure	
	Design and Operation of a MOD 3 Counter	
	Design and Operation of a MOD 5 Counter	
	MOD 6, 7, 9, and 10 Counters	
	Mod 6 and Mod 10 by Cascading Counters	
	TTL Integrated Counters	
9-5	Pseudo Random Counters	260
	Design Procedure	
	Design Example # 1	
	The Design	
	The Operation	
	Design Example # 2	
	The Design	
9-6	Summary Points	267
9-7	Chapter Progress Evaluation	268

Chapter 10 Digital and Analog Conversion

10-1 Introduction 269
10-2 Digital-to-Analog (D/A) Applications 269
10-3 Digital-to-Analog Hardware 271
    WRN
    R2R
10-4 Testing a DAC 276
    Static Test
    Monotonicity Test
10-5 Analog-to-Digital (A/D) Applications 279
10-6 Analog-to-Digital Hardware 281
    Successive-Approximation ADC
    Integration (or Ramp-Type) ADC
    Voltage-to-Frequency (V/F) ADCs
    Counter and Servo ADCs
    Parallel ADC
    Sample-and-Hold Methods
10-7 Summary Points 290
10-8 Chapter Progress Evaluation 291

Chapter 11 Memory

11-1 Introduction 293
11-2 Read Only Memory (ROM) 296
    Diode ROMS
        Operation of Any ROM
        Internal Operation of the Diode Matrix ROM
    Bipolar Transistor ROMs
    MOS FET ROMS
    ROM ICs
11-3 Read Mostly Memory (RMM) 309
    PROMS
    EPROMS
    EEPROM
11-4 Read Write Memory (RWM) 315
    Static Bipolar TTL RAM
        Static TTL RAM Read Operation
        Static TTL RAM Write Operation
    Static MOS RAM
    Dynamic RAM
11-5 Expanding Word Size 327

11-6 Other Types of Solid State Memories . . . 329
    Charge Coupled Device (CCD)
    Josephson Junction Memories
11-7 Magnetic Memory Storage . . . 330
    Magnetic Core Memory
    Magnetic Bubble Memory (MBM)
    Magnetic Tape Storage
    Floppy Disk Storage
11-8 Summary Points . . . 335
11-9 Chapter Progress Evaluation . . . 336

Chapter 12   Introduction to Computers

12-1 Introduction . . . 337
12-2 Arithmetic Logic Unit . . . 339
    Logic Mode
    Arithmetic Mode
12-3 Introduction to the 6802 Microprocessor . . . 348
    Block Diagram of the 6802 MPU
        ALU
        CCR
        Temporary A&B Data Registers
        AccA, AccB
        Instruction Code Register
        Instruction Decoder
        Control Logic Circuitry
        ID
        SP
        Address Register
        6802 MPU Control Lines
        Enable
        $R/\overline{W}$
        VMA
        $\overline{MR}$
        $\overline{Reset}$
        BA
        $\overline{HALT}$
        $\overline{IRQ}$
        NMI
    Program Operation Example
        LDAA

    ADDA
    SWI
12-4 6802 Computer System   356
    MPU
    XTAL
    Power on Reset
    Address Decoder Logic
    User RAM
    Operating System
    PIA
    ACIA
    Memory Map
12-5 Summary Points   360
12-6 Chapter Progress Evaluation   361

Index   363

Chapter 1

Digital Concepts

Objectives

Upon completion of this chapter, you should be able to do the following:

- Describe the characteristics of both analog and digital signals

- List typical types of analog and digital systems

- Define the following basic digital electronic terminology: digital electronics, logic level, positive logic, negative logic, digital gates, truth table, and Boolean algebra

- List the major advantages and disadvantages of analog and digital systems

1-1 Introduction

It is a well-known fact that most electronic systems designed and developed today have sections containing digital circuitry. Digital IC chips are continuously being upgraded to provide better performance with added functions and capabilities. As a result of this rapid development, it becomes necessary for every student of electronics to learn the fundamentals of digital circuitry. Within the computer industry alone, there is a tremendous usage of digital IC chips. When you also consider the new applications of this technology appearing every day, the technician would be severely limited without an understanding of digital electronics.

In addition, you will find the application of digital circuits, in many ways easier than the application of its analog counterpart.

1-2 Analog Signals

Before discussing digital concepts, a well defined knowledge of analog representation and analog signals is needed. First, *analog representation* is defined as any two items that will react in the same fashion but use different unrelated quantities for the comparison. As an example, voltage is an analog representation of water

Digital Circuits and Devices

pressure. Here, voltage is measured in volts and water pressure is measured in pounds per square inch. Notice that a comparison is made using unrelated quantities. As a result, even though the quantities or types of measurements are different, both systems will react in a proportional fashion. In other words, an increase in voltage is analogous to an increase in water pressure.

Analog signals are defined as signals (voltage or current) which vary continuously or smoothly as a function of time. As a result, an analog signal will contain no abrupt steps in level. A simple example would be a DC voltage or current. More detailed examples of analog signals are shown in **Figure 1-1**.

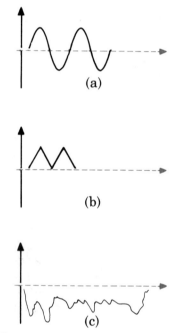

Figure 1-1. Examples of analog signals

Figure 1-1a shows a sine wave that has a voltage level which goes above and below ground changing at a sine function of time as indicated by the horizontal axis. This type of analog signal is the most common.

Figure 1-1b shows a triangle waveform that has a voltage level that stays above ground and changes at a linear function of time.

Figure 1-1c shows a complex waveform that has a voltage level that never goes above ground and changes at a non-linear function of time.

There are many more analog signals than the ones just given. As the definition states, any signal that changes continuously or smoothly is considered an analog signal.

1
Digital Concepts

1-3 Analog Systems

An *analog system* is defined as a device, or group of devices, that use analog signals to perform a particular function and then provide the result in an analog format.

Figure 1-2. Analog Multimeter

One example of an analog system is the analog multimeter shown in **Figure 1-2**. When the voltage on the input changes, the current passing through the meter movement will change proportionally. The change of current through the meter movement will cause the needle to change its position. The measured value is then read off the scale by observing where the needle points. In the case of the analog multimeter, the input signal is constantly being measured, and as the input signal level changes, the position of the needle will also change.

Other examples of analog systems are:

>audio sine-wave generators
>oscilloscopes
>stereo amplifiers
>speakers
>light dimmer switches
>TVs
>light meters.

The systems just listed are normally thought of as analog in nature, because both the input signals and the output representation of the signals are analog. However, with the advent of low cost digital devices, some of these analog systems may also contain digital sections.

Digital Circuits and Devices

1-4 Digital Signals

A *digital signal* is a signal (voltage or current) which normally changes in steps or pulses between two different levels. As with analog signals, digital signal levels can be above, below, or above and below ground.

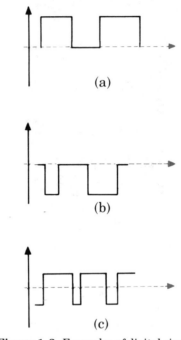

Figure 1-3. Examples of digital signals

Figure 1-3a shows the most common type of digital signal. This type makes up about 85% of the digital signals used today.

Shown in **Figure 1-3b** is the second most commonly used type of digital signal. This "negative" only signal makes up about 10% of the digital signals used. The least used digital signal used today is the one that has levels above and below ground. This type is shown in **Figure 1-3c**. It makes up about the remaining 5% of digital signals.

The common factor in each of these digital signal examples is that only two levels are used or recognized. The most positive level is normally referred to as a *high* logic level (true or 1), while the most negative level is referred to as a *low* logic level (false or 0).

1-5 Digital Systems

A *digital system* is a device, or group of devices that use digital signals to per-

1 Digital Concepts

form a particular function and then display the information in a digital format.

Figure 1-4. Digital Multimeter

Figure 1-4 shows a digital multimeter. When using the multimeter on the volt scale, the input which is an analog signal, is first converted into digital pulses. These pulses are then applied to a digital counter. The output from the counter is then applied to a display driver decoder which will display the data in a numerical character format. Since the results of the measurement are in a character format, the errors that result from reading scales on an analog meter are not present in digital metering.

Other types of digital systems are:
- digital watches
- light switches
- frequency counters
- calculators
- computers
- heart rate monitors
- printers and typewriters.

1-6 Digital Terminology

Digital electronics is the study of electronic switching circuits with its applica-

Digital Circuits and Devices

tion to electronic technology.

Logic level refers to the level on the input or output of a digital device. There are two distinct levels which are identified as a logic *high* level and a logic *low* level. In addition to the two different levels, there are also two different types of logic. The most common type of logic is called ***positive logic***. As shown in **Figure 1-5**, positive logic identifies the most positive voltage level as a logic one or *high*. ***Negative logic*** is shown in **Figure 1-6** where the most negative voltage level represents a logic one or high. Always assume positive logic unless it is specifically mentioned that negative logic is being used.

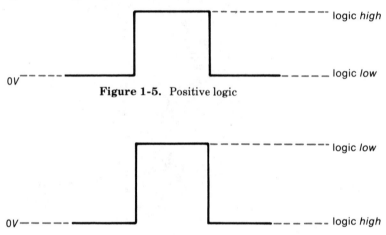

Figure 1-5. Positive logic

Figure 1-6. Negative logic

A *digital gate* is an electronic device that performs a logical decision. The three basic digital gates that make up all digital circuitry are the AND, OR, and NOT gate.

The AND gate is a digital device that will produce a *high* logic level at its output only when all of its input logic levels are *high*. An AND gate can have two or more inputs and only one output. The symbol for a 2-input AND gate is shown in **Figure 1-7**.

Figure 1-7. Symbol for a 2-input AND gate

The OR gate is a digital device which will produce a *high* logic level at its output when any of its input logic levels are *high*. An OR gate can have two or more inputs and only one output. The symbol for a 2-input OR gate is shown in **Figure 1-8**.

1
Digital Concepts

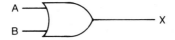

Figure 1-8. Symbol for a 2-input OR gate

The NOT gate (also called an inverter) is a digital device that will produce the opposite logic level at its output than the logic level at its input. As shown in **Figure 1-9**, the NOT gate or inverter has only one input and one output.

Figure 1-9. Symbol for a NOT gate

The internal structure and operation of each of these gates will be discussed in greater detail in the following chapters.

A *truth table* is a diagram using *high* and *low* logic levels which illustrates the output condition of a digital circuit for each possible input combination. The truth tables for the AND, OR, and NOT gate are shown respectively in **Figure 1-10a, b,** and **c**.

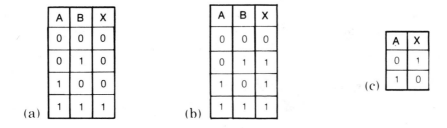

Figure 1-10. Examples of truth tables

1-7 Boolean Algebra

Mathematical operations used to represent the relationship between the input conditions and the output results of a logical decision are called Boolean algebra.

7

Digital Circuits and Devices

This means of representing a logical decision was developed by George Boole, a famous nineteenth century English mathematician and logician.

For the 2-input AND gate shown in **Figure 1-7**, the Boolean expression is $X = A \cdot B$ or $X = AB$. The AND of logic levels is the same as for ordinary multiplication when considering the truth table and Boolean expression at the same time. For the 2-input OR gate of **Figure 1-8**, the Boolean expression is $X = A + B$.

For the NOT or inverter gate shown in **Figure 1-9**, the Boolean expression is $X = \overline{A}$ where the output X is simply the inversion of the input logic level identified as A. Notice that the line or bar above the A is a notation for *inversion* and it is referred to as a *negation line*.

1-8 Digital Vs. Analog

In the example of the analog multimeter, the voltage is used to cause current to flow through the meter movement, which causes the needle to change its position on the scale. In an analog system, the signal itself is used directly to display the results in an analog format.

In the example of the digital multimeter, the voltage applied to the input is converted into a digital signal (pulses). This allows a digital counter to count to a certain value that corresponds to the amount of voltage applied to the input. The count is then converted to a code that will show the measured voltage as a numerical character on the display. In a digital system, the signal is converted into a train of pulses. The pulses are then counted and converted into a digital display character.

Although both the analog and digital multimeter perform the same function, it will be shown why digital systems are being used more frequently. Therefore, let us look at some of the general advantages and disadvantages of analog and digital systems.

1-9 Analog and Digital Systems: Advantages and Disadvantages

The advantages of analog systems are:

- *faster response* since an analog system normally reacts faster to changes in the input signal
- *less distortion* of the input signal since an analog system requires no analog-to-digital conversion which adds distortion to the input signal.

The disadvantages of analog systems are:

- *higher cost, larger size, and more weight* as a result of many different parts to

1
Digital Concepts

perform a specific function
- *more power dissipation* due to the fact that analog systems are normally biased at higher DC voltages and currents which result in a higher consumption of power
- *more power supply levels* since an analog system usually requires more than one power supply level
- *lower display resolution* due to the less accurate display methods of analog systems
- *more complex design* due to the fact that formulas and design procedures are more involved in analog systems
- *more sensitive to operational changes* since small changes in supply levels and operational temperature will cause small changes in the analog signals.

The advantages of digital systems are:
- *lower cost* due to the low cost of IC gates and the ability to put many gates on one IC chip
- *smaller size* since many gates can be put on a single IC chip which allows less IC chips to perform a specific function
- *less weight* due to the lower number of parts for a digital system to perform a specific function
- *less power dissipation* because of the fact that digital systems are biased at lower DC voltages and currents
- *less power supply levels* since a digital system normally requires only one supply level
- *higher display resolution* because digital displays are easier to read
- *easier to design* since a digital system usually consists of various IC chips configured to perform a specific function
- *less sensitive to operational changes* since digital systems use only two levels to determine the type of data and since small changes in supply levels and temperatures normally do not effect operational characteristics.

The disadvantages of digital systems are:
- *slower response* which occurs whenever a digital system converts an analog signal to a digital format due to the time required to make the conversion
- *induced distortion* which occurs whenever an analog input signal is converted to a digital format.

As can be seen from the list of advantages and disadvantages, digital devices are easier to use and design. Along with lower cost and the ability to run off batteries, digital equipment is becoming very popular. Also the price of digital systems will continue to decrease as the number of gates per IC chip is increased.

Digital Circuits and Devices

1-10 Summary Points

1. An analog signal is a signal that varies continuously or smoothly as a function of time without any abrupt steps in level. An analog system consists of a device, or group of devices, that use analog signals to perform a particular function and then provide the result in an analog format.

2. A digital system consists of a device or group of devices that use digital signals to perform a particular function and then display the information in digital format. A digital gate is a device that performs a logical decision.

3. Positive logic is the most common type of logic where the most positive voltage level represents a logic *high* or one, and the most negative voltage level represents a logic *low* or zero.

4. A truth table is a diagram using *high* and *low* logic levels which illustrates the output condition of a digital circuit for each possible input combination.

5. Boolean algebra refers to the mathematical operations used to represent the relationship between the input conditions and the output results of a logical decision.

1-11 Chapter Progress Evaluation

1. A signal that changes in steps or pulses of which only two levels are recognized is called a/an?

2. Digital signals only have levels that are positive. True or false?

3. What are the three basic gates that make up all digital circuits?

4. Which gate will produce a *high* logic level at its output when any one or more of its inputs is at a *high* logic level? (assume positive logic)

5. Which one of the following Boolean expressions represents a 2-input AND gate with A and B as its inputs: $X = A+B$, $X = AB$, or $X = \overline{AB}$?

Chapter 2

Numbering Systems

Objectives

Upon completion of this chapter, you should be able to do the following:

- Describe the characteristics of binary, hexadecimal, and octal
- Count to any numerical value in binary, hexadecimal, and octal
- Convert values between any two numbering systems
- Describe the characteristics of the 8421 and excess-3 BCD codes
- Describe the basic characteristics of Gray code, ASCII, and EBCDIC coding systems

2-1 Introduction

Binary means having two distinct parts. In the binary numbering system, it means then that there are two different characters that are valid in any one digit position. Since there are only two different characters or digits in binary, it is called a *base 2* number system. Binary values consist of numbers made up from the two digits 1 and 0.

In base 2 numbering, the highest single-digit value is 1; therefore, only a 1 or a 0 can occupy any digit position. All digital devices perform yes/no or true/false decision-making operations. To represent the choices, binary values 1 and 0 are used. These binary values are referred to as *bits*. In general, 1 stands for yes, while 0 stands for no.

The binary number system is a very important numbering system since computers and digital electronics operate in binary. Even when other numbering systems and codes are entered into a computer, the computer must first convert the value into binary before any other operation can be performed. Therefore, a good understanding of the binary system is needed for a technician to work on computers and digital equipment.

Digital Circuits and Devices

COUNTING IN BINARY

In any numbering system, the count starts with zero and increments by 1 in the LSD (Least Significant Digit) or LSB (Least Significant Bit when talking about computers) until it reaches the highest single-digit value allowed by the number system. Incrementing again then turns the LSD to 0, and a carry increments the next most significant digit.

When counting in binary, the number is read "ones" and "zeros." For example in **Table 2-1**, the first binary count is read "zero," the second binary count is read "one," the third is read "one-zero," and so forth. The only time "one-zero" is pronounced ten is in the decimal (base 10) number system. The same goes for hundreds, thousands, and so on.

Table 2-1.
Binary Counting

Decimal	Binary
0	0
1	1
2	10
3	11
4	100
5	101
6	110
7	111
8	1000
9	1001
10	1010
11	1011
12	1100
13	1101
14	1110
15	1111
16	1 0000
17	1 0001

DIGIT POSITION WEIGHTS FOR DECIMAL AND BINARY

Any decimal number can be broken down into smaller values. For example, the decimal number 1111 is equal to the sum of $1000 + 100 + 10 + 1$. Notice that each digit position carries a specific *weight*. For the decimal number system, the weight of each digit position is found by raising 10 to the specific digit position as shown below in **Figure 2-1**.

digit position	etc.	2	1	0	−1	−2	etc.
weight	etc.	$10^2 = 100$	$10^1 = 10$	$10^0 = 1$	$10^{-1} = 1/10$	$10^{-2} = 1/100$	etc.

decimal point

Figure 2-1. Decimal weights

2
Numbering Systems

Notice that each digit position to the left of the decimal point has a weight ten times more than the previous digit position, whereas each digit position to the right of the decimal point has a weight ten times less than the previous digit position. As an example, the decimal number 35.8 consists of 3 weights of ten, 5 weights of one, and 8 weights of one-tenth.

It can be shown for any number system that the weight of a specific digit position is equal to the base of the specific number system raised to the specific digit position. As a result, **Figure 2-2** illustrates the digit position weights for the binary number system which are also called bit position weights.

digit position	etc.	3	2	1	0	−1	−2	etc.
weight	etc.	$2^3 = 8$	$2^2 = 4$	$2^1 = 2$	$2^0 = 1$	$2^{-1} = 1/2$	$2^{-2} = 1/4$	etc.

binary point

Figure 2-2. Binary weights

In the binary number system, each digit position to the left of the binary point has a weight twice that of the previous digit position, whereas each digit position to the right of the binary point has a weight of half that of the previous digit. For example, the binary number 101.1 consists of 1 weight of four, 0 weight of two, 1 weight of one, and 1 weight of one-half.

The binary weights of digit positions 0 through 15, and the corresponding decimal values are listed in **Table 2-2**.

Table 2-2.
Powers of 2 for Binary to Decimal Conversion

Digit Position	Binary Weights	Decimal Equivalent
0	2^0	1
1	2^1	2
2	2^2	4
3	2^3	8
4	2^4	16
5	2^5	32
6	2^6	64
7	2^7	128
8	2^8	256
9	2^9	512
10	2^{10}	1024
11	2^{11}	2048
12	2^{12}	4096
13	2^{13}	8192
14	2^{14}	16384
15	2^{15}	32768

Digital Circuits and Devices

BINARY TO DECIMAL CONVERSION

Converting a binary number to its equivalent decimal value involves 3 steps:

1. Label the digit or bit positions 0, 1, 2, 3, etc. to the left of the binary point and −1, −2, −3, etc. to the right of the binary point if necessary.
2. Calculate the binary weight of each digit position that contains a 1 by raising 2 to the digit position (the bit positions containing a 0 have no decimal value since zero times any binary weight is zero).
3. Add the resulting binary weights to determine the equivalent decimal number.

For example:

$$\begin{array}{rl} \text{Digit Position} & 4\ 3\ 2\ 1\ 0 \\ \text{Binary Number} & 1\ 0\ 1\ 1\ 0 \end{array}$$

$$\begin{aligned} 2^1 &= 2 \\ 2^2 &= 4 \\ 2^4 &= \underline{16} \\ & 22 \end{aligned}$$

Therefore, binary 10110 is equal to decimal 22.

DECIMAL TO BINARY CONVERSION

There are two methods of converting decimal values to binary. One is called *progressive subtraction*; the other is known as *modulo division*.

In progressive subtraction, use **Table 2-2** while following these steps:

1. Find the largest power-of-2 value that can be subtracted from the decimal value to be converted. Its power-of-2 represents the position of the most significant bit in the binary value that will result. Put a 1 at that bit position.
2. Now subtract. The difference becomes the next value from which a power-of-2 must be subtracted.
3. Repeat 1 and 2 until the decimal value is reduced to zero.
4. Enter a 0 in all bit positions where a 1 was not placed.

Figure 2-3 provides an example of this conversion technique. The decimal number to be converted to binary is 7370. The 12th power-of-2, decimal 4096, is the highest power that can be taken from 7370, so a 1 goes at bit 12. Subtracting 4096 leaves 3274. The 11th power-of-2 can be subtracted from 3274, so another 1 goes at bit 11.

Continuing these subtractions, the 10th power-of-2 is less than 1226. The 9th and 8th powers-of-2 are too large for 202, but the 7th power is okay. Then the 6th power-of-2 works for 74, the 3rd fits into 10, and, finally, the 1st can be taken from 2.

Numbering Systems

As a result of these operations, you should place a binary 1 at bits, 12, 11, 10, 7, 6, 3, and 1. Then applying step 4, fill the remaining bit positions with binary 0, and just like that decimal 7370 is converted to binary 1110011001010. If you should question the accuracy of your work, convert the binary back to decimal as a means of proof.

```
DECIMAL TO BE CONVERTED                    7 3 7 0
LARGEST POWER OF 2      (2^12)           − 4 0 9 6
                                           3 2 7 4
LARGEST POWER OF 2      (2^11)           − 2 0 4 8
                                           1 2 2 6
                        (2^10)           − 1 0 2 4
                                             2 0 2
                        (2^7)            −   1 2 8
                                                7 4
                        (2^6)            −     6 4
                                                1 0
                        (2^3)            −       8
                                                  2
                        (2^1)            −       2
                                                  0
BINARY PATTERN FORMED (BITS)
12 11 10 9 8 7 6 5 4 3 2 1 0
 1  1  1 0 0 1 1 0 0 1 0 1 0
```

Figure 2-3. Decimal to binary conversion using progressive subtraction

In *modulo division*, you divide repeatedly by the base value while saving only the remainders. Here are the steps to follow:

1. Divide the decimal value by 2 (the base).
2. Write down the remainder. This is the least significant bit of the equivalent binary number.
3. Divide the quotient thus derived by 2 again.
4. Again write down the remainder. This is the next bit in the equivalent binary number.
5. Repeat steps 3 and 4 until the quotient equals zero.

Figure 2-4 illustrates this method by showing the conversion of decimal 2936 to its equivalent binary value. Applying step 1 results in a quotient of 1468 with a remainder of 0. This remainder becomes the LSB of the equivalent binary number. Steps 3 and 4 are repeated until the quotient equals zero. The last division which yields a remainder of 1 becomes the Most Significant Bit (MSB) of the equivalent binary number. As a result, decimal 2936 is equivalent to binary 1011 0111 1000.

Digital Circuits and Devices

```
2 9 3 6 / 2 =   1468    R = 0    LSB
1 4 6 8 / 2 =    734    R = 0
  7 3 4 / 2 =    367    R = 0
  3 6 7 / 2 =    183    R = 1
  1 8 3 / 2 =     91    R = 1
    9 1 / 2 =     45    R = 1
    4 5 / 2 =     22    R = 1
    2 2 / 2 =     11    R = 0
    1 1 / 2 =      5    R = 1
      5 / 2 =      2    R = 1
      2 / 2 =      1    R = 0
      1 / 2 =      0    R = 1    MSB

RESULT:          1 0 1 1  0 1 1 1  1 0 0 0
```

Figure 2-4. Decimal to binary conversion using modulo division

Conversion from decimal to binary using modulo division is easy. Any decimal value that ends in an even digit leaves a 0 remainder, while a value ending with an odd digit forces a 1 remainder.

Topic Review   2-1

1. The term _____ means having two distinct parts.

2. Binary 00011011 is equivalent to decimal _____ .

3. Binary 110.001 is equivalent to decimal _____ .

4. _____ _____ and _____ _____ are two methods of converting decimal values to binary.

5. Decimal 268 is equivalent to binary _____ .

Answers:

1. binary
2. 27
3. 6.125
4. Progressive subtraction, modulo division
5. 100001100

2-2 Hexadecimal Numbering System

HEXADECIMAL SYSTEM

The word *hex* means a value of six while the root word *deci* means one-tenth. Therefore, the word *hexadecimal* refers to a number system that can contain up to sixteen different characters or symbols per digit position. The hexadecimal number system is known as a *base 16* numbering system or *hex* for short. The importance of a base 16 number system will be seen in a later section on converting between the hexadecimal and binary number systems.

COUNTING IN HEX

The first ten characters comprising the hexadecimal number system are the ten characters of the decimal number system (0, 1, 2, 3, 4, 5, 6, 7, 8, 9). The remaining six characters are taken from the first six letters of the alphabet (A, B, C, D, E, F). This sixteen character set which makes up the hexadecimal number system is tabulated in **Table 2-3**. This table shows the hexadecimal characters or single digits and their corresponding decimal and binary equivalents.

Table 2-3.
Character Set for the Hexadecimal Number System

Binary	Decimal	Hexadecimal
0000	0	0
0001	1	1
0010	2	2
0011	3	3
0100	4	4
0101	5	5
0110	6	6
0111	7	7
1000	8	8
1001	9	9
1010	10	A
1011	11	B
1100	12	C
1101	13	D
1110	14	E
1111	15	F

Counting in hexadecimal is the same as in any other number system. Simply follow the general procedure outlined below.

1. Start with zero.
2. Increment one step at a time, until you reach the highest digit the base allows.

Digital Circuits and Devices

3. Increment once again. The least significant digit (LSD) reverts to zero and sends a carry of 1 to the next more significant digit, incrementing it.
4. Increment one step at a time until both digit positions contain the highest digit value the base allows.
5. Increment again. This turns the LSD to zero and forces a carry, which turns the next significant digit position to zero and forces another carry.
6. Continue adding a more significant digit whenever all current digits reach the highest value allowed.

Now let us apply these general principles of counting to the hexadecimal number system. Starting with the first hex digit 0, increment one step at a time until the highest single digit allowed for in base 16 is reached. **Table 2-3** shows the result up to this point. Upon incrementing again, the first digit position reverts to 0, and the second digit position increments to 1 as a result of the carry of 1 to the next significant digit. Further counting increments the LSD again until the highest character allowed for is reached. Incrementing again, reverts the first digit position back to 0, and the carry of 1 to the next significant digit increments that digit position. Continuing this procedure results in the following counting in hex:

0, 1, 2, 3, 4, 5, 6, 7, 8, 9, A, B, C, D, E, F, 10, 11, 12, 13, 14, 15, 16, 17, 18, 19, 1A, 1B, 1C, 1D, 1E, 1F, 20, ..., 2F, 30, ..., 3F, 40, ..., FF, 100, and so on.

When counting in hex, it is important to overcome the habit of calling the second row "ten, eleven, twelve, thirteen,...." since that manner of counting applies to the decimal number system. Notice that after counting past hexadecimal 19, there is no decimal term to fall back on. In addition to this, hexadecimal 20 is not equal to decimal 20. As a result when counting in hex, use the terms "one-zero, one-one, one-two, one-three, ...;" it will then be much easier to continue: "..., one-nine, one-ay, one-bee, one-cee, one-dee, one-eee, one-eff, ... etc." Also when conversing with another engineer or technician, they will realize immediately that you are referring to counting in hex, rather than decimal.

Hexadecimal numbering or counting in hex will play a key part in all your future studies of digital electronics and microprocessor systems. If you have any doubts about your understanding of counting in base 16, continue to practice until you become familiar with the pattern of hexadecimal counting.

HEX/BINARY CONVERSION

Notice how difficult it is to read, recognize, decipher, and use large binary numbers such as 1101111101111010. Although digital circuits and microcomputer systems have no trouble reading, interpreting, or operating upon huge bit patterns, technicians and engineers need a more recognizable system of notation. The base 16 number system meets this need since any binary number that consists of a large number of bits can be broken down into groups of four bits each. A group of four bits is commonly referred to as a *nibble*, whereas a group of eight bits (equivalent to two nibbles) is commonly called a *byte*. Notice that each nibble (equivalent to a half-byte), whatever its bit pattern, can be represented by a hexadecimal digit. Furthermore, an entire byte requires only two hex digits. Application of this concept makes it possible for large binary numbers to be represented as equivalent hexadecimal numbers which are much smaller and thereby easier for technicians to use. As an example, the binary number 00011011 looks very similar to binary 00011001, and it is only when these two binary numbers are next to each other that they appear distinctly different in their bit pattern. It can be shown that the equivalent values of these binary numbers are hexadecimal 1B and 19, respectively. Notice how the equivalent hexadecimal numbers are much easier to distinguish and remember. Also, since most computers use eight or sixteen bits of data to perform various operations, every byte of data (eight bits) can be represented using only two hex digits.

Converting a binary number to its equivalent hexadecimal value involves only two steps:

1. Break down the binary number into groups of four bits starting at the binary point and going in both directions if necessary.
2. Convert each nibble or group of four bits to its hex digit using **Table 2-3**.

In a similar manner, converting in the opposite direction from hex to binary again involves only two steps:

1. Using **Table 2-3** write down the four-bit binary equivalent for each hexadecimal number.
2. For easier recognition, leave the equivalent binary value in four-bit groups.

Given the steps to convert from binary to hex and vise versa, notice how easy it is in the following examples to convert between these two number systems:

$$00011011_2 = 0001\ 1011_2 = 1B_{16}$$

$$19_{16} = 0001\ 1001_2$$

The use of *subscript notation* in the above examples is the way mathematicians often indicate the base of a number. That is why hexadecimal 19 is written 19_{16} which means "to the base 16." In a similar manner, decimal 19 using subscript notation is written 19_{10}. Subscript notation can be extended to any number system. Notice how binary 11_2 is easily distinguished from decimal 11_{10}.

Instead of using subscript notation, computer programmers use *program-symbol notation* to indicate the base of a numeric value. For example the prefixed dollar sign ($) is used as a common symbol for hexadecimal. In other words, hexadecimal 1B

Digital Circuits and Devices

would be identified as $1B. Binary numbers are sometimes prefixed with a percent symbol, as in %00011011.

Both subscript and program-symbol notation are often used as means of representing the base of a number. In general, subscripts are best suited for mathematical usage, and symbols apply best to programming.

HEX TO DECIMAL CONVERSION

Table 2-4 lists the hexadecimal weights of digit positions 0 through 9 along with the corresponding decimal values.

Table 2-4.
Power of 16 for Hex to Decimal Conversion

Digit Position	Hex Weight	Decimal Value
0	16^0	1
1	16^1	16
2	16^2	256
3	16^3	4096
4	16^4	65 536
5	16^5	1 048 576
6	16^6	16 777 216
7	16^7	268 435 456
8	16^8	4 294 967 296
9	16^9	68 719 476 736

This table will be used to convert hex values to decimal as outlined below:

1. Label the digit positions 0, 1, 2, 3, etc. to the left of the hexadecimal point.
2. Calculate the decimal value of each hex digit by multiplying the hex weight of each position by the decimal equivalent of the hex value occupying that digit position.
3. Add the result of each multiplication to determine the equivalent decimal number.

For example:

Digit Position 3 2 1 0
Hex Number B 9 2 A

$A_{16} = 10_{10}$ $16^0 \times 10_{10} = 10$
$2_{16} = 2_{10}$ $16^1 \times 2_{10} = 32$
$9_{16} = 9_{10}$ $16^2 \times 9_{10} = 2304$
$B_{16} = 11_{10}$ $16^3 \times 11_{10} = \underline{45056}$
47402

Therefore, hex B92A is equal to decimal 47402.

20

2 Numbering Systems

DECIMAL TO HEX CONVERSION

Decimal to hex conversion can be accomplished by either progressive subtraction or by modulo division. However, for large decimal values, modulo division works best. The steps to follow for modulo 16 division are:

1. Divide the decimal value by 16 (the base).
2. Write down the decimal remainder and convert it to a hexadecimal digit. This value forms the least significant digit of the target hexadecimal value.
3. Divide the quotient thus derived by 16.
4. Again, write down the decimal remainder and convert it to hex. This is the next digit for the hex value.
5. Repeat steps 3 and 4 until the quotient equals zero.

The final remainder, expressed in hex notation, becomes the most significant digit of the converted hexadecimal value.

An example of using modulo division to convert from decimal to hex is given below in **Figure 2-5**. Notice that dividing decimal 45178 by 16 gives a quotient of 2823 and a remainder of 10. The hex equivalent of decimal 10 is A_{16} which becomes the LSD of the hexadecimal value being found.

Dividing the quotient 2823 by 16 produces a new quotient of 176. The remainder is 7, which is also 7 in hexadecimal. This provides the second digit for the hex value.

Next, divide 176 by 16. The answer is 11, and the remainder is 0, since 176 divides evenly by 16. The third digit in the hex value is 0.

Dividing 11 by 16, you get a quotient of 0 — with 11 remaining. Hexadecimal B is equivalent to decimal 11, so B is the most significant digit of the hexadecimal value. Therefore, decimal 45178 converts to hexadecimal B07A.

```
45178 / 16 = 2823     R = 10 = A    LSD
 2823 / 16 =  176     R =  7 = 7
  176 / 16 =   11     R =  0 = 0
   11 / 16 =    0     R = 11 = B    MSD
```
Result:
Decimal 45178 = Hexadecimal B07A

Figure 2-5. Decimal to hex conversion using modulo division

Topic Review 2-2

1. A group of four bits is commonly referred to as a _____, whereas a group of eight bits is referred to as a _____.

Digital Circuits and Devices

2. Binary 11011010 is equivalent to hex _____ .
3. Binary _____ is equivalent to hex F4.

4. Hex 6A is equivalent to decimal _____ .

5. Hex _____ is equivalent to decimal 77.

Answers:

1. nibble, byte
2. DA
3. 11110100
4. 106
5. 4D

2-3 Octal Numbering System

COUNTING IN OCTAL

The *octal* numbering system contains eight different characters or symbols, and it is referred to as the *base 8* numbering system. These eight digits are the first eight characters of the decimal number system (0, 1, 2, 3, 4, 5, 6, 7).

Applying the general principles of counting to the octal number system results in the following counting in octal:

0, 1, 2, 3, 4, 5, 6, 7, 10, 11, 12, 13, 14, 15, 16, 17, 20, 21, 22, and so on.

As with the other nondecimal numbers, it helps to clarify the octal value by saying "one-zero, one-one, one-two, ..."

Using subscript notation, octal 377 is written 377_8, and it is read as "three-seven-seven to the base 8". For program-symbol notation, octal numbers are shown with an "at" symbol (@) as a prefix. For example, octal 377 would be identified as @377.

OCTAL/BINARY CONVERSION

The conversion from octal to binary and vice versa is centered around the relationship between their character sets. Remember, that the binary or base 2 character set consists of 0 and 1, while the octal or base 8 character set consists of 0, 1, 2, 3, 4, 5, 6, and 7. **Table 2-5** shows the basic relationship between octal and binary characters or digits.

2
Numbering Systems

Table 2-5.
Basic Relationship Between Octal and Binary Digits

Octal	Binary
0	000
1	001
2	010
3	011
4	100
5	101
6	110
7	111

Converting a binary number to its equivalent octal value involves just two steps:

1. Break the binary number into groups of three bits starting at the binary point and going in both directions if necessary.
2. Convert each group of three bits to its octal equivalent using **Table 2-5**.

For example:

$$0111 1010_2 = 001\ 111\ 010_2 = 172_8$$
$$101.0111_2 = 101.011\ 100_2 = 5.34_8$$

Converting an octal number to its equivalent binary value is accomplished by Converting each octal digit to its binary equivalent using **Table 2-5**.

For example:

$$377_8 = 011\ 111\ 111_2 = 11\ 111\ 111_2$$

OCTAL/HEX CONVERSION

Conversion from octal to hex and vice versa is quick and easy since the binary number system is common between both octal and hex.

Converting an octal number to hex involves three steps:

1. Convert each digit of the octal number to its binary equivalent using **Table 2-5**.
2. Break the resulting binary number into groups of four bits starting at the binary point and going in both directions if necessary.
3. Convert each group of four bits to its hexadecimal equivalent using **Table 2-3**.

For example:

$$074_8 = 000\ 111\ 100_2 = 0000\ 0011\ 1100_2 = 03C_{16} = 3C_{16}$$
$$15.3_8 = 001\ 101.011_2 = 0000\ 1101.0110_2 = 0D.6_{16} = D.6_{16}$$

23

Digital Circuits and Devices

Converting a hexadecimal number to its equivalent octal value involves three steps:

1. Convert each hex digit to its binary equivalent using **Table 2-3**.
2. Break the resulting binary number into groups of three bits starting at the binary point and going in both directions if necessary.
3. Convert each group of three bits to its octal equivalent using **Table 2-5**.

For example:

$$FA_{16} = 1111\ 1010_2 = 011\ 111\ 010_2 = 372_8$$
$$E.7_{16} = 1110.0111_2 = 001\ 110.011\ 100_2 = 16.34_8$$

OCTAL TO DECIMAL CONVERSION

Table 2-6 lists the octal weights of digit positions 0 through 8 along with the corresponding decimal values.

Table 2-6.
Power of 8 for Octal to Decimal Conversion

Digit Position	Octal Weight	Decimal Value
0	8^0	1
1	8^1	8
2	8^2	64
3	8^3	512
4	8^4	4096
5	8^5	32 768
6	8^6	262 144
7	8^7	2 097 152
8	8^8	16 777 216

This table will be used to convert octal values to decimal as outlined below:

1. Label the digit positions to the left of the octal point as 0, 1, 2, 3, etc.
2. Calculate the decimal value of each octal digit be multiplying the octal weight of each digit position by the decimal equivalent of the octal value occupying that digit position.
3. Add the result of each multiplication to determine the equivalent decimal number.

For example:

    Digit Position   2 1 0
    Octal Number   2 5 6

2
Numbering Systems

$$6_8 = 6_{10} \qquad 8^0 \times 6_{10} = 6$$
$$5_8 = 5_{10} \qquad 8^1 \times 5_{10} = 40$$
$$2_8 = 2_{10} \qquad 8^2 \times 2_{10} = \underline{128}$$

Therefore, $256_8 = 174_{10}$.

DECIMAL TO OCTAL CONVERSION

Any decimal number can be converted to any other number system by either progressive subtraction or by modulo division. Modulo division involves less work especially when converting large decimal numbers. The steps to follow for modulo 8 division are:

1. Divide the decimal value by 8.
2. Write down the decimal remainder and convert it to an octal digit. This value forms the LSD of the target octal value.
3. Divide the quotient thus derived by 8.
4. Write down the decimal remainder and convert it to octal. This is the next digit for the target octal value.
5. Repeat steps 3 and 4 until the quotient equals zero.

For example:

$$\text{Decimal } 47168 = \text{Octal?}$$

$47168/8 =$	5896	$R = 0$	$= 0$	LSD
$5896/8 =$	737	$R = 0$	$= 0$	
$737/8 =$	92	$R = 1_{10}$	$= 1_8$	
$92/8 =$	11	$R = 4_{10}$	$= 4_8$	
$11/8 =$	1	$R = 3_{10}$	$= 3_8$	
$1/8 =$	0	$R = 1_{10}$	$= 1_8$	MSD

Therefore, $47168_{10} = 134100_8$.

SPLIT-OCTAL NOTATION

Microcomputers often deal with two-byte instructions and data. Hexadecimal notation is well-suited to handle binary values of this size. The four-digit hex value $C07F_{16}$ consists of sixteen bits since each hex digit represents four bits. In other words, $C0_{16}$ and $7F_{16}$ represent one-byte values where their binary or bit patterns are 1100 0000 and 0111 1111, respectively. Therefore, the bit pattern of $C07F_{16}$ is 1100 0000 0111 1111.

Combining bytes in octal cannot be done as easy as in hex. In fact, octal notation

25

Digital Circuits and Devices

is best suited for one-byte values. However, two-byte octal values can be written in a manner call *split-octal notation*. The only purpose of this notation is to show two-byte octal values. For example, the octal bytes 377_8 and 203_8 can be shown as the two-byte octal number 377_8-203_8. In forming the binary equivalent of this two-byte octal number, the most significant octal digit of each byte generates two bits just as it would if the bytes were alone.

Topic Review 2-3

1. Binary 10111011 is equivalent to octal _____ .

2. Octal 371 is equivalent to hex _____ .

3. Hex AD is equivalent to octal _____ .

4. Octal 52 is equivalent to decimal _____ .

5. Decimal 85 is equivalent to octal _____ .

Answers:

1. 273
2. F9
3. 255
4. 42
5. 125

2-4 Binary-Coded Decimal Numbers

BCD CODES

Decimal values remain the most familiar way to digitally display measurements. However, digital equipment can work or operate only in a binary format. As a result, a means of digitally converting from binary to decimal is needed. This need has resulted in the development of many *binary coded decimal* (BCD) codes. In BCD codes, the digits of a decimal number are encoded into groups of binary digits. Converting from decimal to BCD code is called *encoding*, while converting from BCD to decimal is called *decoding*.

2 Numbering Systems

THE 8421 CODE

The most natural BCD code is the *8421 code* which is a weighted code that expresses the decimal digits by their four-bit binary equivalents. A *weighted* binary code is one where each bit position has a fixed value.

Table 2-7 illustrates the relationship between the 8421 code and the single-digits or characters of the decimal number system. Notice that each single-digit decimal number is represented by an equivalent four-bit group.

Table 2-7.
Basic Relationship Between Decimal and 8421 Code Digits

Decimal	8421
0	0000
1	0001
2	0010
3	0011
4	0100
5	0101
6	0110
7	0111
8	1000
9	1001

Using **Table 2-7**, any decimal number can be converted to its equivalent 8421 code value and vice versa.

For example:

$$597_{10} = 0101\ 1001\ 0111 \text{ in 8421 code}$$
$$0001\ 0011 \text{ in 8421 code} = 13_{10}$$

In other words, decimal 597 can be encoded in 8421 code as the bit pattern 0101 1001 0111, while the bit pattern 0001 0011 in 8421 code can be decoded to decimal 13.

Although the 8421 code doesn't provide fewer bits, the following advantages outweigh the size handicap of the 8421 coded bit patterns:

1. They are easily translated into decimal equivalents.
2. They can represent analog values in digital form.
3. They are grouped in nibbles (groups of four bits) which are easy for byte-oriented digital and computer systems to work with.

THE EXCESS-3 CODE

Another widely used BCD code is the *excess-3 code*. This method of encoding a

Digital Circuits and Devices

BCD number is obtained by adding 3_{10} to each decimal digit and then converting each result to binary. **Table 2-8** results from this operation where it shows the relationship between the excess-3 code, and the single digits of the decimal number system. As in the 8421 code, each single digit decimal number is represented in the excess-3 code by an equivalent four bit group. Using **Table 2-8**, any decimal number can be encoded to or decoded from excess-3 code.

Table 2-8.
Basic Relationship Between Decimal and Excess-3 Code Digits

Decimal	Excess-3
0	0011
1	0100
2	0101
3	0110
4	0111
5	1000
6	1001
7	1010
8	1011
9	1100

The excess-3 code is an unweighted code since each bit position has no fixed value. In other words, it is not possible to assign weights or values to the various bit positions.

Topic Review   2-4

1. _____ _____ _____ codes provide a way to encode the digits of a decimal number into groups of binary digits.

2. The decimal number 97 is encoded as _____ in 8421 code.

3. The _____ code is a weighted BCD code, whereas the _____ code is an unweighted BCD code.

Answers:

1. Binary Coded Decimal
2. 1001 0111
3. 8421, excess-3

2-5 Coding Systems

GRAY CODE

The *Gray code* is an unweighted binary code which is used in computer communications. In Gray code, only one binary bit changes from one count to the next sequential count. Since only one bit will change between sequential counts, the chance of error in communication will decrease to only one bit. An unweighted binary code means that each bit position has no fixed value. As a result, Gray code is not suited for arithmetic operations. The Gray code is shown in **Table 2-9** along with its corresponding decimal and binary numbers.

Table 2-9.
Gray Code up to 15_{10}

Decimal	Binary	Gray Code
0	0000	0000
1	0001	0001
2	0010	0011
3	0011	0010
4	0100	0110
5	0101	0111
6	0110	0101
7	0111	0100
8	1000	1100
9	1001	1101
10	1010	1111
11	1011	1110
12	1100	1010
13	1101	1011
14	1110	1001
15	1111	1000

ASCII

ASCII stands for American Standard Code of Information Interchange. The ASCII code uses seven binary bits to represent the English alphabet, numbers, punctuation symbols, and printer control codes. The ASCII code system is the most used code in today's computer world. It is used in about 75% of all communications between computers. The complete ASCII code is shown in **Table 2-10**.

Digital Circuits and Devices

Table 2-10
ASCII Character Set

LSD \ MSD	0 000	1 001	2 010	3 011	4 100	5 101	6 110	7 111
0 0000	NUL	DLE	SPACE	0	@	P		p
1 0001	SOH	DC1	!	1	A	Q	a	p
2 0010	STX	DC2	"	2	B	R	b	r
3 0011	ETX	DC3	#	3	C	S	c	s
4 0100	EOT	DC4	$	4	D	T	d	t
5 0101	ENG	NAK	%	5	E	U	e	u
6 0110	ACK	SYN	&	6	F	V	j	v
7 0111	BEL	ETB	'	7	G	W	f	w
8 1001	BS	CAN	(	8	H	X	g	x
9 1001	HT	EM	)	9	I	Y	h	y
A 1010	LF	SUB	*	:	J	Z	i	z
B 1011	VT	ESC	+	;	K	[	k	{
C 1100	FF	FS	,	<	L	\	l	|
D 1101	CR	GS	-	=	M	]	m	}
E 1110	SO	RS	.	>	N	↑	n	~
F 1111	SI	VS	/	?	O	←	o	DEL

Table 2-11
The EBCDIC Chart

MSB	0	1	2	3	4	5	6	7	8	9	A	B	C	D	E	F
0	NUL				SP	&	-									0
1	SDH	DC1					/		a	j		A		J		1
2	STX	DC2	SYN						b	k	s		B	K	S	2
3	ETX	DC3							c	l	t		C	L	T	3
4		RES	BYP						d	m	u		D	M	U	4
5	HT	NL	LF	DC4					e	n	v		E	N	V	5
6		BS	EOB						f	o	w		F	O	W	6
7	DEL			EOT					g	p	x		G	P	X	7
8		CAN							h	q	y		H	Q	Y	8
9		EN							i	r	z		I	R	Z	9
A						¢	!		:							
B	VT					.	$	,	#	{	}					
C	FF	FLS					<	*	%	@						
D	CR	GS	ENQ	NAK	(		)		-	,						
E	SO	RDS	ACK		+		;	>	=			[			]	
F	SI	US	BEL	SUB			?	"								
LSB																

2
Numbering Systems

EBCDIC

EBCDIC stands for Extended Binary Coded Decimal Interchange Code. EBCDIC is an eight-bit IBM binary code used to represent the English alphabet, numbers, punctuation symbols, and printer control codes. EBCDIC is the second most used code in computer communications today.

By looking at the EBCDIC chart shown in **Table 2-11** you will notice that there are many codes which do not represent any characters. IBM uses the blank code locations for special characters that are not used for every computer.

2-6 Summary Points

1. The decimal numbering system contains ten characters, and it is referred to as the base 10 numbering system.

2. The binary or base 2 numbering system contains 2 fundamental characters.

3. The importance of the binary number system is due to the fact that computers and digital electronics only understand binary bit patterns.

4. The hexadecimal or base 16 numbering system contains 16 fundamental characters.

5. A nibble consists of a group of four bits, while a group of eight bits is commonly called a byte.

6. Large binary numbers can be represented as equivalent hexadecimal numbers which are much smaller and easier to use.

7. Conversion from decimal to any other number system can be accomplished by either progressive subtraction or by modulo division.

8. The octal or base 8 numbering system contains 8 fundamental characters or digits.

9. Octal notation is best suited for one-byte values; however, two-byte octal values can be written in a manner called split-octal notation.

10. Binary coded decimal codes provide a way to encode the digits of a decimal number into groups of binary digits.

Digital Circuits and Devices

11. The 8421 code is a weighted BCD code, whereas the excess-3 code is an unweighted BCD code.

12. An unweighted binary code means that each bit position has no fixed value.

13. The Gray code is an unweighted binary code that is used in computer communications.

2-7 Chapter Progress Evaluation

Convert the following numbers into the proper system:

1. 85_{10} = _____ $_2$

    = _____ $_8$

    = _____ $_{16}$

2. $0001\ 1100_2$ = _____ $_{10}$

    = _____ $_8$

    = _____ $_{16}$

3. 5252_8 = _____ $_{10}$

    = _____ $_2$

    = _____ $_{16}$

4. $8AD_{16}$ = _____ $_{10}$

    = _____ $_2$

    = _____ $_8$

5. EBCDIC $4E_{16}$ = _____ (CHARACTER)

6. EBCDIC $F9_{16}$ = _____ (CHARACTER)

Chapter 3

Digital Gates

Objectives

Upon completion of this chapter, you should be able to do the following:

* Explain the basic characteristics of an AND, OR, Inverter, NAND, NOR, ex-OR, and ex-NOR gate

* Draw a diode and switch circuit which will function as an AND and OR gate

* Draw a transistor circuit which will function as an AND, OR, NOT, NAND, and NOR gate

* Draw a gate circuit which will function as an ex-OR and ex-NOR gate

* Derive a truth table, Boolean expression, and timing diagram for the various

* Define sign, sense, and sign & sense gate inversion

3-1 Introduction

Digital circuitry is composed of building blocks called gates and other functional circuits such as memory cells, flip-flops, and gate combinations. A gate is simply a device having one output and one or more inputs such that the output state is determined by the state of the input or inputs. In the vast majority of cases, digital circuitry is made up various combinations of gates. In fact, even the flip-flop is made up of gate combinations.

In this chapter we will examine the functional properties of the basic gates used as building blocks of digital circuitry. The most common gates are the AND gate, OR gate, and logic inverter, sometimes called a NOT gate. Combinations of these gates make up other types of gates, but by and large it is safe to say that these three gates

Digital Circuits and Devices

make up the foundation for building digital logic circuitry.

You will also see in this chapter that schematic diagrams of digital logic gates are much easier to work with than the conventional circuit diagrams you are used to using. They contain circuit components in a "building block" format rather than in discrete form and present themselves in a more "viewable" fashion making it easier to follow signal flow through the various sections of the digital circuitry.

3-2 The AND Gate

The AND gate is defined as a digital device that will produce a *high* on its output only when all inputs are *high*. AND gates have two or more inputs and only one output. The circuit shown in **Figure 3-1a** shows the representation of a 2-input AND gate function using two SPST switches. Point X is the output of the AND gate and the LED is used to give a visual indication of the output level. The 470 ohm resistor in series with the LED limits the current when the LED is on. The LED (Light Emitting Diode) will not turn on until the voltage across it reaches approximately 2VDC; in the AND gate circuit the LED will either be on or off.

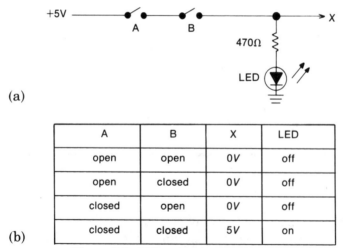

A	B	X	LED
open	open	0V	off
open	closed	0V	off
closed	open	0V	off
closed	closed	5V	on

Figure 3-1. AND gate operational representation and truth table

In **Figure 3-1a,** when switch A and B are open, no current will flow through the circuit, the voltage at point X will be zero and the LED will be off. When either switch A or switch B is closed, there will still be no current flowing through the circuit, the voltage at point X will again be zero and the LED is off. Finally, when switch A and B are both closed, current will flow in the circuit, the voltage at point X will be approximately 5VDC, and the LED will light.

3
Digital Gates

The truth table is a diagram which will demonstrate how the circuit works with all possible input conditions.

There are two types of truth tables: one is called an internal operational truth table, and is shown in **Figure 3-1b**; the other is called a logic truth table or just truth table. The internal operational truth table gives the conditions of the inputs, output and internal parts of the circuit; normally this type of truth table is not given because the conditions of the internal parts are not necessary when the circuit is in an integrated circuit. Note that the internal operational truth table of a gate circuit is important to first understand the operation of the gate. The most popular type of truth table is the *logic truth table* or just truth table. This type, shown in **Table 3-1**, deals with the inputs and output of the circuit. When reviewing data sheets the logic truth table is the only one given to explain the operation of the gate.

Table 3-1.
AND gate logic truth table

A	B	X
0	0	0
0	1	0
1	0	0
1	1	1

As can be seen from both truth tables the only time the output goes *high* is when all inputs are *high;* at all other times the output will be *low.*

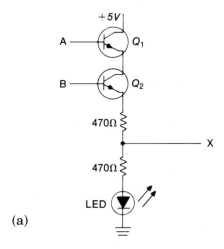

(a)

Figure 3-2. Transistor AND gate and operational truth table

35

Digital Circuits and Devices

(b)

A	B	Q_1	Q_2	X	LED
0V	0V	off	off	0V	off
0V	5V	off	on	0V	off
5V	0V	on	off	0V	off
5V	5V	on	on	3V	on

Figure 3-2. (continued)

Figure 3-2a shows a two-input AND gate function using transistors. The transistors act as *switches;* they turn on and conduct current when forward biased. Again the operational truth table is shown in **Figure 3-2b.** Because this is still a two-input AND gate, the logic truth table shown in **Table 3-1** applies.

The Boolean expression for any gate is a mathematical expression which will represent how the gate will function. The following is the Boolean expression for the 2-input AND gate.

$$X = A \cdot B \quad \text{or} \quad X = AB$$

Where the (·) multiplication sign or the lack of a symbol with no space between two or more of the variables represent the AND function. The Boolean expression can be read two different ways: X is equal to A AND B, or X is equal to AB ANDed.

The circuit shown in **Figure 3-3a**, is a 3-input AND gate function using diodes. The internal operational truth table will again explain the operation of the circuit.

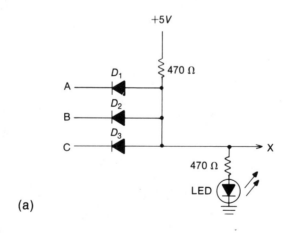

(a)

Figure 3-3. 3-input diode AND gate

A	B	C	D_1	D_2	D_3	X	LED
0V	0V	0V	on	on	on	.6V	
0V	0V	5V	on	on	off	.6V	
0V	5V	0V	on	off	on	.6V	
0V	5V	5V	on	off	off	.6V	
5V	0V	0V	off	on	on	.6V	
5V	0V	5V	off	on	off	.6V	
5V	5V	0V	off	off	on	.6V	
5V	5V	5V	off	off	off	V+	

(b)

Figure 3-3. (continued)

As can be seen from the internal operational truth table shown in **Figure 3-3b**, when any of the diodes are on, there will be approximately .6V at point X, which is the value of one diode voltage drop. Only when all diodes are off will the voltage at point X be high enough to allow the LED to turn on. The voltage at point X will be approximately V+ minus the voltage drop across the 470 ohm resistor.

Table 3-2.
3-input AND gate logic truth table

A	B	C	X
0	0	0	0
0	0	1	0
0	1	0	0
0	1	1	0
1	0	0	0
1	0	1	0
1	1	0	0
1	1	1	1

The Boolean expression for the 3-input AND gate is:

$$X = A \cdot B \cdot C \quad \text{or} \quad X = ABC$$

Digital Circuits and Devices

A schematic (logic) symbol like those shown in Figure 3-4 is an abbreviated way of showing the function of a gate without drawing all the internal parts of the gate.

The output is always taken from the curved end of the symbol. The inputs of any AND gate are always on the flat side of the symbol.

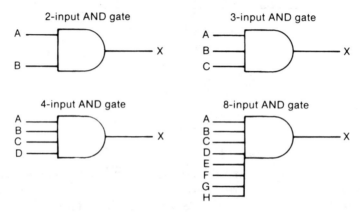

Figure 3-4. AND gate schematic symbols

Note the symbols just given are the most common sizes of AND gates. Note also on the 8-input AND gate there is a line that extends beyond the symbol; this is sometimes done in order to keep the size of each gate the same, and still allow all inputs to be shown.

TIMING DIAGRAM

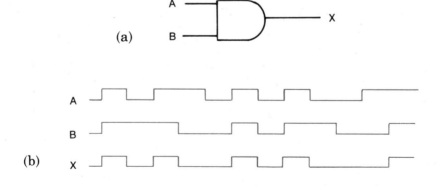

Figure 3-5. AND gate timing diagram

3
Digital Gates

A timing diagram is an illustration which shows the output and input conditions of a gate circuit over a given period of time; see **Figure 3-5.** The output of the circuit in the timing diagram is determined by the input conditions at any given time.

For example, in order to determine the output (X), look at the two inputs of **Figure 3-5a** and plug their logic levels into the AND gate logic truth table. It can be seen that whenever both inputs are *high* the output will be *high* during that time period. At all other times the output will be *low*.

As can be seen from the timing diagram of AND gates, the only time the output will be *high* is when all inputs are *high*. At all other times the output will be *low*.

INCREASING INPUTS

Increasing the number of inputs an AND gate circuit can have is very easy, all that has to be done is to, AND, AND gates together. The following are circuits and truth tables to show how to increase the number of inputs in an AND gate.

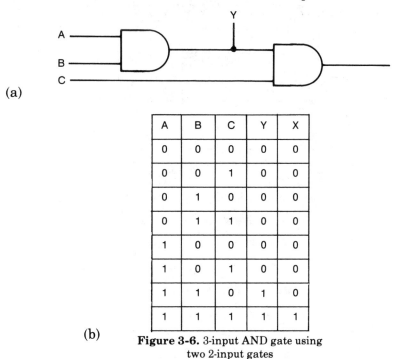

Figure 3-6. 3-input AND gate using two 2-input gates

By disregarding point Y in **Figure 3-6b**, which is just used to help determine the output of the last AND gate, the truth table is the same as a 3-input AND gate.

Digital Circuits and Devices

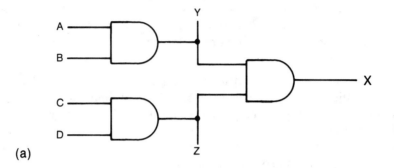

A	B	C	D	Y	Z	X
0	0	0	0	0	0	0
0	0	0	1	0	0	0
0	0	1	0	0	0	0
0	0	1	1	0	1	0
0	1	0	0	0	0	0
0	1	0	1	0	0	0
0	1	1	0	0	0	0
0	1	1	1	0	1	0
1	0	0	0	0	0	0
1	0	0	1	0	0	0
1	0	1	0	0	0	0
1	0	1	1	0	1	0
1	1	0	0	1	0	0
1	1	0	1	1	0	0
1	1	1	0	1	0	0
1	1	1	1	1	1	1

(a)
(b)

Figure 3-7. 4-input AND gate using three-input gates

By disregarding points Y and Z you will note that the truth table just given is the truth table for a 4-input AND gate.

3
Digital Gates

The circuit shown in **Figure 3-8a** is another way of getting a 4-input AND gate function using three 2-input AND gates.

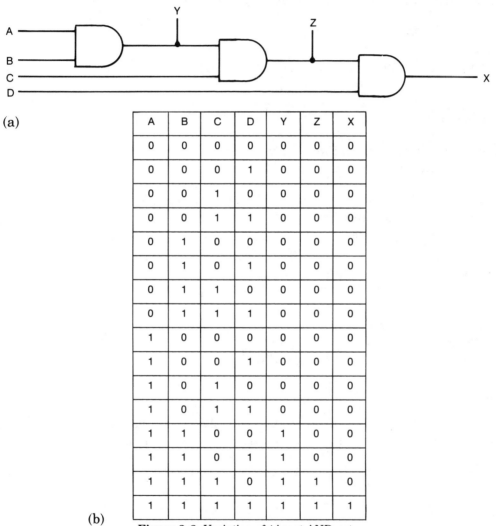

A	B	C	D	Y	Z	X
0	0	0	0	0	0	0
0	0	0	1	0	0	0
0	0	1	0	0	0	0
0	0	1	1	0	0	0
0	1	0	0	0	0	0
0	1	0	1	0	0	0
0	1	1	0	0	0	0
0	1	1	1	0	0	0
1	0	0	0	0	0	0
1	0	0	1	0	0	0
1	0	1	0	0	0	0
1	0	1	1	0	0	0
1	1	0	0	1	0	0
1	1	0	1	1	0	0
1	1	1	0	1	1	0
1	1	1	1	1	1	1

Figure 3-8. Variation of 4-input AND gate

By disregarding point Y and Z, you will note that this is also a truth table for a 4-input AND gate. Either method will work for getting a 4-input AND function from three 2-input AND gates.

As can be seen from the previous examples, in order to increase the number of inputs an AND gate function can have, you simply AND, AND gates together.

41

Digital Circuits and Devices

Topic Review 3-2

1. The AND gate will produce a *high* output when all its inputs are _____.

2. The two different truth tables are called _____ and _____.

3. The _____ table is given when dealing with the inputs and output condition of a gate.

4. The Boolean expression for the 2-input AND gate is _____.

5. A timing diagram is an illustration which shows the output and input conditions of a _____ circuit over a given _____ of _____.

6. To increase the number of inputs an AND gate circuit can have, you just _____ the AND gates together.

Answers:
1. *high*
2. operational, logic
3. truth
4. X = AB or X = A · B
5. gate, period, time
6. AND

3-3 The OR Gate

The OR gate is defined as a digital device that will produce a *high* on its output, when any of its inputs are *high*. OR gates have two or more inputs and only one output.

The circuit shown in **Figure 3-9a** shows the representation of a 2-input OR gate function using two SPST switches.

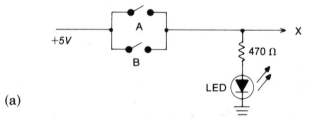

(a)

Figure 3-9. OR gate operational representation and truth table

3
Digital Gates

A	B	X	LED
open	open	0V	off
open	closed	5V	on
closed	open	5V	on
closed	closed	5V	on

(b)

Figure 3-9. (continued)

In **Figure 3-9a**, A and B are SPST switches, which are used as the two inputs for the circuit. Point X is the output of the OR gate and the LED is used to represent the output state. When the LED is off, the logic level of point X is *low;* when the LED is on, the logic level of point X is *high*.

As can be seen from the internal operational truth table in **Figure 3-9b,** the only time the output of the gate is *low* is when all switches are open (low), at any other time the output will be *high*.

The logic truth table shown in **Table 3-3** uses only logic levels on the inputs and output to show the operation of the circuit. In the logic truth table, when the switch is open it represents a logic 0, and when the switch is closed, it represents a logic 1. The output is represented by a logic 0 when point X is about 0V, and when point X is $\frac{1}{2}$ supply or more it is represented by a logic 1.

Table 3-3.
OR gate logic truth table

A	B	X
0	0	0
0	1	1
1	0	1
1	1	1

The circuit shown in **Figure 3-10** shows a 2-input OR gate function using transistors. Once again, the transistors take the place of the switches.

The 3 volts at point X when the output is *high* is due to the voltage drop across the transistor and series resistor.

The Boolean expression for a two input OR gate with A and B as inputs and X as the output is as follows:

$$X = A + B$$

43

Digital Circuits and Devices

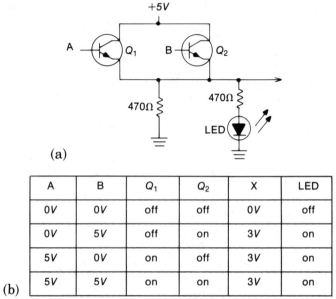

(a)

A	B	Q_1	Q_2	X	LED
0V	0V	off	off	0V	off
0V	5V	off	on	3V	on
5V	0V	on	off	3V	on
5V	5V	on	on	3V	on

(b)

Figure 3-10. Transistor OR gate and operational truth table

The plus sign (+) in Boolean is referred to as an OR sign. Therefore X = A + B is read, X is equal to A OR B; another way of reading the expression is X is equal to A, B ORed.

The circuit shown in **Figure 3-11a** shows a 3-input OR gate function, using diodes. The internal operational truth table shown in **Figure 3-11b** will explain the operation of the circuit.

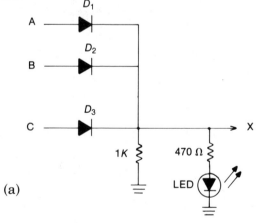

(a)

Figure 3-11. 3-input diode OR gate

44

A	B	C	D_1	D_2	D_3	X	LED
0V	0V	0V	off	off	off	0V	off
0V	0V	5V	off	off	on	V+	on
0V	5V	0V	off	on	off	V+	on
0V	5V	5V	off	on	on	V+	on
5V	0V	0V	on	off	off	V+	on
5V	0V	5V	on	off	on	V+	on
5V	5V	0V	on	on	off	V+	on
5V	5V	5V	on	on	on	V+	on

(b)

Figure 3-11. (continued)

As can be seen from the internal operational truth table the only time the output will be *low* is when all inputs are *low* and all diodes are off; in this condition no current will flow through the LED and point X will be at ground potential. Any time a *high* is placed on an input to the OR gate, the diode for that input will be forward biased; this will place V+ at point X and will allow current to flow through the LED.

Table 3-4.
3-input OR gate logic truth table

A	B	C	X
0	0	0	0
0	0	1	1
0	1	0	1
0	1	1	1
1	0	0	1
1	0	1	1
1	1	0	1
1	1	1	1

The Boolean expression for the 3-input OR gate is:

$$X = A + B + C$$

Digital Circuits and Devices

The schematic symbol (logic symbol) is an abbreviated way of showing a function of a digital gate, without drawing all internal parts of the gate circuit. The most common OR gate symbols are shown in **Figure 3-12**.

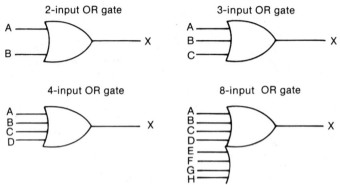

Figure 3-12. OR gate schematic symbols

TIMING DIAGRAM

Remember, timing diagrams are illustrations which show the output condition of a gate, for any type of input condition. The timing diagram for the 2-input and 3-input OR gate is shown in **Figure 3-13**.

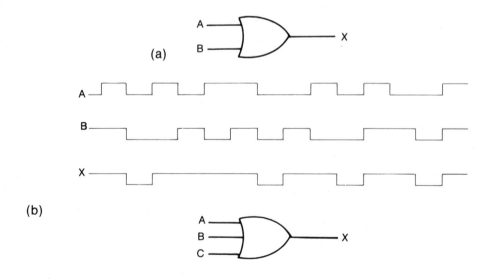

Figure 3-13. OR gate timing diagrams

3
Digital Gates

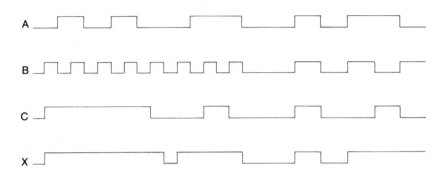

Figure 3-13. (continued)

INCREASING INPUTS

Increasing the number of inputs of an OR gate is very simple; all that has to be done is to OR the output of one OR gate to another, as shown in **Figure 3-14a** and **Figure 3-15a**.

(a)
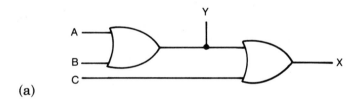

(b)

A	B	C	Y	X
0	0	0	0	0
0	0	1	0	1
0	1	0	1	1
0	1	1	1	1
1	0	0	1	1
1	0	1	1	1
1	1	0	1	1
1	1	1	1	1

Figure 3-14. 3-input OR gate using two 2-input gates

47

Digital Circuits and Devices

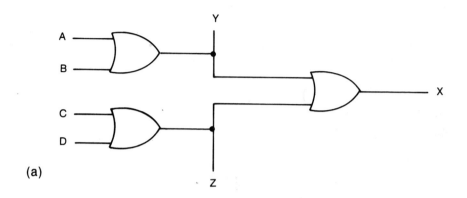

A	B	C	D	Y	Z	X
0	0	0	0	0	0	0
0	0	0	1	0	1	1
0	0	1	0	0	1	1
0	0	1	1	0	1	1
0	1	0	0	1	0	1
0	1	0	1	1	1	1
0	1	1	0	1	1	1
0	1	1	1	1	1	1
1	0	0	0	1	0	1
1	0	0	1	1	1	1
1	0	1	0	1	1	1
1	0	1	1	1	1	1
1	1	0	0	1	0	1
1	1	0	1	1	1	1
1	1	1	0	1	1	1
1	1	1	1	1	1	1

(b)

Figure 3-15. 4-input OR gate using three 2-input OR gates

By disregarding points Y and Z the truth table looks like a standard 4-input OR gate truth table.

Digital Gates

Topic Review 3-3

1. The OR gate will produce a *high* output when any of its inputs are _____.

2. The Boolean expression for the 2-input OR gate is _____.

3. To increase the number of inputs an OR gate can have, you just _____ the output of one OR gate to another.

Answers:

1. *high*
2. X = A + B
3. OR

3-4 Inverter

The digital inverter (or sometimes called the NOT gate) is defined as a digital device that will output the opposite condition than the condition that is on its input. Inverters have only one input and only one output. The circuit shown in **Figure 3-16a** demonstrates an inverter function using a transistor.

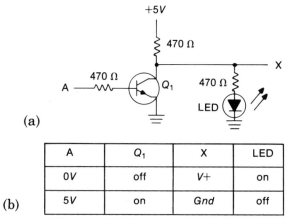

A	Q_1	X	LED
0V	off	V+	on
5V	on	Gnd	off

(a)

(b)

Figure 3-16. Transistor inverter (NOT) gate and operational truth table

In the inverter circuit, the transistor is used as a switch. With zero volts on the base of Q_1, the transistor will be turned off and the voltage at point X will be approximately V+, minus the voltage drop across the collector resistor. When 5V or a *high* is applied to the base of Q_1, the transistor will go into saturation and the voltage at

Digital Circuits and Devices

point X will be approximately 1V, this is the voltage dropped across the saturated transistor. **Table 3-5** shows the logic truth table.

Table 3-5.
Inverter (NOT) logic truth table

A	X
0	1
1	0

As can be seen from the truth table the output of the gate will always be the opposite condition than the input condition.

The Boolean expression for the inverter is:

$$X = \overline{A}$$

The expression is read X is equal to A NOT or X is equal to NOT A.

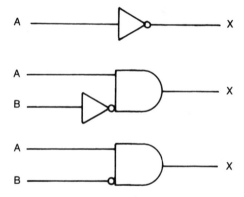

Figure 3-17. The inverter schematic symbol

The inverter logic (schematic) symbol can be broken down into two parts. The triangle section less the circle (bubble) is commonly used to signify a buffer amplifier. However, by adding the circle on the output, the symbol becomes the designator for an inverter. As shown in **Figure 3-17a**, the input is always on the buffer and the output is taken from the point opposite the input (bubble). When the inverter is used on either input of a gate, it may be illustrated as shown in **Figure 3-17b** or **3-17c**.

TIMING DIAGRAM

As can be seen from the timing diagram in **Figure 3-18**, the inverter output is always the opposite condition from the level on its input.

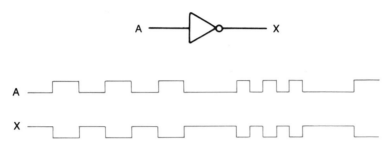
Figure 3-18. Inverter (NOT) timing diagram

Topic Review   3-4

1. The inverter is sometimes called a _____ gate.

2. The output condition of an inverter is always _____ to the input.

3. The Boolean expression for the inverter is _____ .

4. The _____ portion of the inverters logic symbol actually signifies inversion.

Answers:

1. NOT
2. opposite
3. $X = \overline{A}$
4. circle or bubble

3-5 The NAND Gate

The word NAND means "NOT AND"; a NAND gate is an AND gate with an inverted output. A NAND gate is defined as a digital device having a *high* on its output when any of its input are *low*. A NAND gate will have two or more inputs and only one output. The truth table of a NAND gate has the opposite condition of the AND gate.

The circuit shown in **Figure 3-19** performs a NAND function using transistors.

As can be seen from the internal operational truth table, as long as any one of the transistors is off, the output at point X will be approximately V+ minus the voltage dropped across the collector resistor. When all the transistors are in satura-

Digital Circuits and Devices

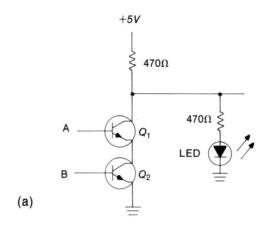

(a)

(b)

A	B	Q_1	Q_2	X	LED
0V	0V	off	off	V+	on
0V	5V	off	on	V+	on
5V	0V	on	off	V+	on
5V	5V	on	on	Gnd	off

Figure 3-19. Transistor NAND gate and operational truth table

tion, the output voltage will be the voltage dropped across the transistors, or about 1V.

Table 3-6 shows the logic truth table for the NAND gate.

Table 3-6.
NAND gate logic truth table

A	B	X
0	0	1
0	1	1
1	0	1
1	1	0

The Boolean expression looks the same as the AND gate expression, except the entire expression is negative (inverted).

3
Digital Gates

$$X = [\overline{AB}] \text{ or } X = [\overline{A \cdot B}]$$

The Boolean expression is read X is equal to A, B NANDed or X is equal to A NAND B. Note the negation sign must be above all the inputs of the NAND gate and also the AND sign to make it a valid NAND Boolean expression.

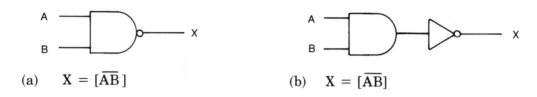

(a) $X = [\overline{AB}]$ (b) $X = [\overline{AB}]$

Figure 3-20. NAND gate schematic symbol and discrete NAND gate

The symbol for the NAND gate shown in **Figure 3-20a** looks like an AND gate with an inverter symbol (circle) on the output; this inverts the output condition before it leaves the integrated circuit. **Figure 3-20b** shows an AND gate with an external inverter. Both circuits will work the same and have the same truth tables, but one circuit has two gates and the other has only one gate. Note also that the negation sign should be within brackets in order to leave no doubt that the example, using the one gate is a NAND gate, not just an AND gate with an inverted output. The NAND gate has the inverter built into the circuit, not external; that is why the negation sign is inside the brackets, not outside the brackets. The brackets shows the output of the gate.

TIMING DIAGRAM

The timing diagrams for NAND gates are shown in **Figure 3-21**. Just as in all other timing diagrams the output will be determined by the input conditions, using the truth table for determining the output.

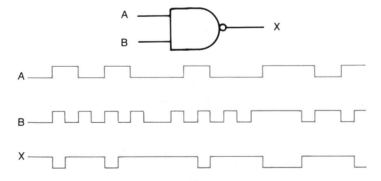

Figure 3-21. NAND gate timing diagrams

53

Digital Circuits and Devices

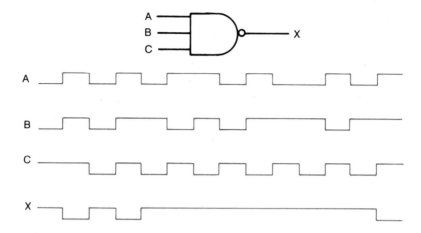

Figure 3-21. (continued)

As can be seen from the timing diagrams, the only time the output of the NAND gate is *low* is when all its inputs are *high*.

INCREASING INPUTS

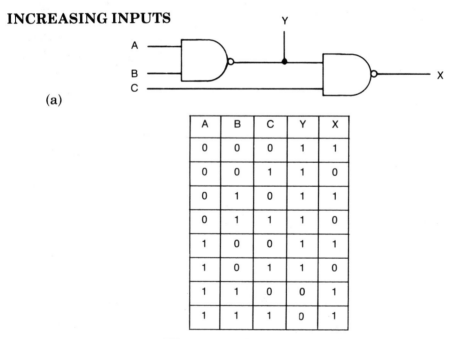

(a)

A	B	C	Y	X
0	0	0	1	1
0	0	1	1	0
0	1	0	1	1
0	1	1	1	0
1	0	0	1	1
1	0	1	1	0
1	1	0	0	1
1	1	1	0	1

(b)

Figure 3-22. Non-Functioning 3-input NAND gate using two 2-input NAND gates

3
Digital Gates

Increasing the number of inputs a NAND gate has is not as simple as with the AND gates studied earlier. The best way, if possible, is to use the proper size NAND gate. If it is not possible, you will have to NAND outputs from AND gates.

By disregarding point Y from the truth table (Figure 3-22b) you will note that this is not the truth table for a 3-input NAND gate; this proves that you cannot increase the number of inputs a NAND gate circuit has by just NANDing NAND gates together.

Figure 3-23a shows the proper way of increasing the number of inputs a NAND gate circuit can have and still operate like a single NAND gate.

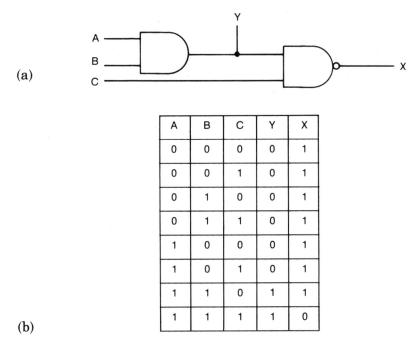

Figure 3-23. Functional 3-input NAND gate using 2-input AND and NAND gates

If point Y is disregarded you will note that this is the truth table for a three input NAND gate. So to increase the number of inputs a NAND circuit can have, just AND additional inputs and then NAND the final outputs from the AND gates.

NAND INVERTER

If an inverter is not available, but an inverter is needed, a NAND gate can be used as an inverter. In order to make a NAND gate inverter, two methods can be

Digital Circuits and Devices

utilized. The best way is to use one of the NAND inputs as the inverter input and take all other NAND inputs to the supply as shown in **Figure 3-24a.** The second method is to tie all NAND inputs together as shown in **Figure 3-24b** and use it as the input of the inverter. The second method is not recommended because it reduces the number of NAND inverters that an output of a gate can drive; this is due to the maximum output current of any gate.

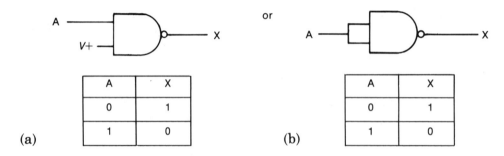

Figure 3-24. NAND inverters

Topic Review   3-5

1. The word NAND means _____ _____.

2. A NAND gate is an AND gate with an _____ output.

3. The Boolean expression for a 2-input NAND gate is _____.

4. To increase the number of inputs a NAND circuit can have, just _____ additional inputs and then _____ the final outputs.

Answers:

1. NOT AND
2. inverted
3. $X = \overline{[AB]}$ or $X = \overline{[A \cdot B]}$
4. AND, NAND

3-6 The NOR Gate

The word NOR means "NOT OR"; a NOR gate is an OR gate with a built-in inverter. A NOR gate is defined as a digital device which will produce a *high* on its output only when all its inputs are *low*. The truth table of a NOR gate has the opposite output condition as the OR gate. A NOR gate has two or more inputs and only one output.

The circuit shown in **Figure 3-25** performs a NOR function using transistors.

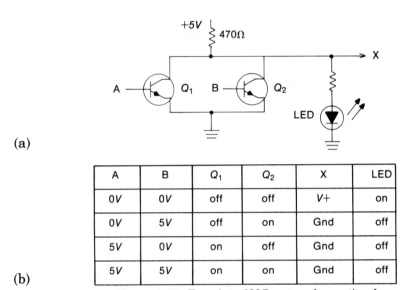

(a)

(b)

A	B	Q_1	Q_2	X	LED
0V	0V	off	off	V+	on
0V	5V	off	on	Gnd	off
5V	0V	on	off	Gnd	off
5V	5V	on	on	Gnd	off

Figure 3-25. Transistor NOR gate and operational truth table

As can be seen from the internal operational truth table, as long as all of the transistors are off the output at point X will be approximately V+ minus the voltage drop across the collector resistor. When any of the transistors are in saturation, the output voltage will be about 1 VDC or the voltage dropped across a saturated transistor.

Table 3-7 shows the logic truth table for the NOR gate.

Table 3-7.
NOR gate logic truth table

A	B	X
0	0	1
0	1	0
1	0	0
1	1	0

Digital Circuits and Devices

The Boolean expression looks the same as the OR gate expression, except the entire expression is negatived (inverted).

$$X = [\overline{A + B}]$$

The Boolean expression is read X is equal to A NOR B or X is equal to AB NORed. Note, the negation sign must be above all the inputs of the expression to make it a valid NOR Boolean expression.

The symbol for the NOR gate is shown in **Figure 3-26a** and looks like an OR gate with an inverter symbol (circle) on the output. This inverts the output condition before it leaves the integrated circuit. **Figure 3-26b** shows a NOR gate with an external inverter. Notice again, the difference in Boolean expressions.

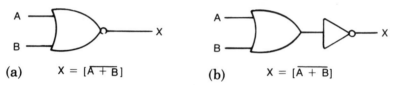

Figure 3-26. NOR gate schematic symbol and discrete NOR gate

When the negation line is inside the brackets the inverter is built inside the gate circuit.

TIMING DIAGRAM

The following timing diagrams shown in **Figure 3-27a** and **3-27b** are for NOR gates. Just as in all other timing diagrams, the output will be determined by the input conditions, using the truth table.

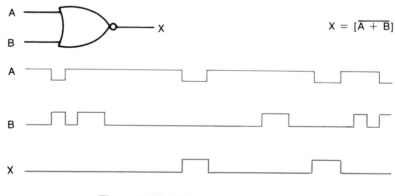

(a)

Figure 3-27. NOR gate timing diagrams

3
Digital Gates

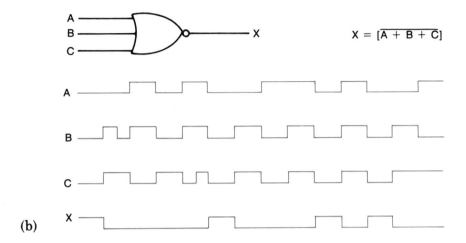

(b)

Figure 3-27. (continued)

As can be seen from the timing diagram, the only time the output of the NOR gate is *high* is when all its inputs are *low*.

INCREASING INPUTS

Increasing the number of inputs a NOR gate has, using small size NOR gates is not as simple as with OR gates. The best way, possibly, is to use the proper size NOR gate. If it is not possible, you will have to NOR outputs from OR gates to achieve the NOR function. The following will prove how to expand the number of inputs a NOR circuit can have.

A	B	C	Y	X
0	0	0	1	0
0	0	1	1	0
0	1	0	0	1
0	1	1	0	0
1	0	0	0	1
1	0	1	0	0
1	1	0	0	1
1	1	1	0	0

Figure 3-28. Non-functioning 3-input NOR gate using two 2-inpt NOR gates

Digital Circuits and Devices

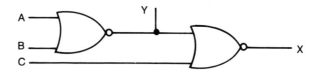

Figure 3-28. (continued)

By disregarding point Y from the truth table, you will note that the truth table is not the truth table for a 3-input NOR gate function; this proves that you cannot increase the number of inputs a NOR gate function has by NORing NOR gates together.

The configuration shown in **Figure 3-29a** is the proper way of increasing the number of inputs of a NOR gate function.

A	B	C	Y	X
0	0	0	0	1
0	0	1	0	0
0	1	0	1	0
0	1	1	1	0
1	0	0	1	0
1	0	1	1	0
1	1	0	1	0
1	1	1	1	0

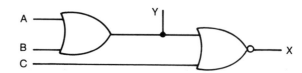

Figure 3-29. Functional 3-input NOR gate using a 2-input OR and NOR gates

If point Y is disregarded, you will note that this is the truth table for a three input NOR gate. So to increase the number of inputs a NOR circuit can have, OR additional inputs and then NOR the final outputs from the OR gates.

NOR INVERTER

If an inverter is not available, but an inverter is needed, a NOR gate can be used.

Digital Gates

In order to make a NOR gate an inverter, two methods can be used. The best way is to use one of the NOR inputs as the inverter input, and then take all other NOR inputs to ground as shown in **Figure 3-30a.** The second method is to tie all NOR inputs together as shown in **Figure 3-30b,** and use it as the input of the inverter. The second method is not recommended because it reduces the number of NOR inverters that can be used in the circuit.

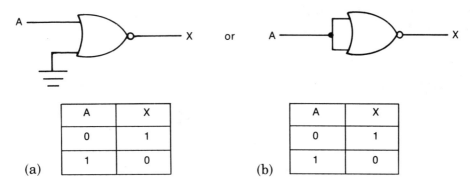

Figure 3-30. NOR inverters

Topic Review 3-6

1. The word NOR means _____ _____ .

2. A NOR gate will produce a *high* output when all its inputs are _____ .

3. The Boolean expression for a NOR gate is _____ .

4. To increase the number of inputs a NOR circuit can have, you _____ additional inputs and then _____ the outputs.

Answers:

1. NOT OR
2. *low*
3. $X = \overline{[A + B]}$
4. OR, NOR

3-7 Exclusive-OR Gate

The definition of an exclusive-OR gate is a digital device that will produce a *high*

Digital Circuits and Devices

on its output, only when its inputs are at different levels. An ex-OR gate has only two inputs and one output.

DISCRETE EX-OR CIRCUIT

The ex-OR gate is a combination of two AND gates, one OR gate and two inverters; they are connected together to achieve the ex-OR function as shown in **Figure 3-31**. This gate is very useful in adder circuits and digital difference circuits, where an indication must be made when the two input levels are different.

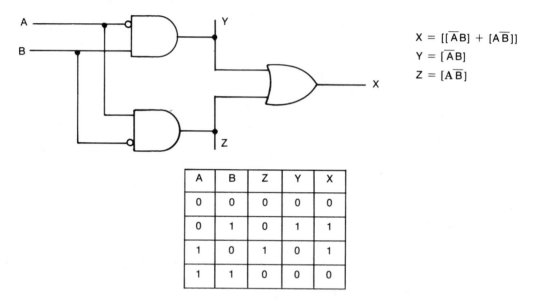

Figure 3-31. Discrete ex-OR circuit, operational truth table and Boolean expression

By disregarding points Y and Z on the internal operational truth table, you will note that the only time the output of the circuit (point X) goes *high* is when the two inputs are different. When the inputs are the same the output will be *low*.

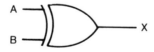

Figure 3-32. Exclusive-OR (ex-OR) logic symbol

The logic truth table for the ex-OR gate is shown in **Table 3-8**.

3
Digital Gates

Table 3-8.
Exclusive-OR gate logic truth table

A	B	X
0	0	0
0	1	1
1	0	1
1	1	0

The logic (schematic) symbol for an integrated circuit ex-OR gate is shown in **Figure 3-32;** and the Boolean expression is as follows:

$$X = [A \oplus B]$$

Where the "$\oplus$" sign is the symbol for a ex-OR gate function, the brackets represents the output of the gate.

TIMING DIAGRAM

The timing diagram for an ex-OR gate is shown in **Figure 3-33.**

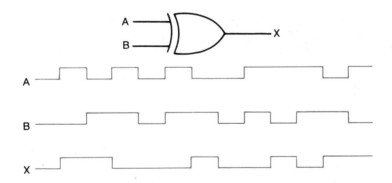

Figure 3-33. Exclusive-OR timing diagram

As can be seen from the timing diagram, the only time the output is *high* is when the two inputs are at different levels.

Topic Review   3-7

1. The definition of an ex-OR gate is a digital device that will produce a *high* on its

Digital Circuits and Devices

output only when its inputs are at _____ levels.

2. The Boolean expression for the ex-OR gate is _____ .

Answers:

1. different
2. $X = [A \oplus B]$

3-8 Exclusive-NOR Gate

The definition for an exclusive-NOR (ex-NOR) gate is a digital device that will produce a *high* on its output, only when all of its inputs are at the same level. An ex-NOR gate has only two inputs and only one output.

DISCRETE EX-NOR CIRCUIT

The ex-NOR gate is a combination of two AND gates, one OR gate and two inverters, which are connected together to achieve the ex-NOR function. This gate is very useful in computer communication circuits, where an indication is needed when two levels are the same. The digital circuit shown in **Figure 3-34** is a discrete circuit for an ex-NOR gate function.

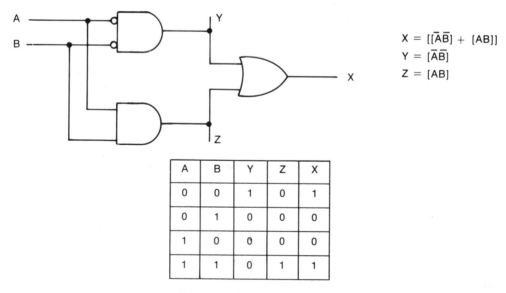

A	B	Y	Z	X
0	0	1	0	1
0	1	0	0	0
1	0	0	0	0
1	1	0	1	1

$X = [[\bar{A}\bar{B}] + [AB]]$
$Y = [\bar{A}\bar{B}]$
$Z = [AB]$

Figure 3-34. Discrete ex-NOR circuit, operational truth table and Boolean expression

3
Digital Gates

By disregarding points Y and Z on the internal operational truth table, you will note that the only time the output of the circuit (point X) goes *high* is when the inputs are the same. When the inputs are different the output will be *low*.

The logic truth table for the ex-NOR gate is shown in **Table 3-9.**

Table 3-9.
Exclusive-NOR gate logic truth table

A	B	X
0	0	1
0	1	0
1	0	0
1	1	1

The logic symbol for an integrated circuit ex-NOR gate is shown in **Figure 3-35.**

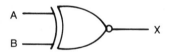

Figure 3-35. Exclusive-NOR gate logic symbol

The double curved input section of the symbol indicates an ex-OR function, the output circle (inverter) on the symbol indicates that the symbol is an ex-NOR gate function. You should also note that the truth table for the ex-NOR has the opposite output condition as does the ex-OR function.

The Boolean expression for the IC ex-NOR gate is:

$$X = \overline{[A \oplus B]}$$

The "$\oplus$" sign is the symbol for the ex-OR gate function, but since the output has a built-in negation sign, the $\oplus$ symbol represents an ex-NOR gate function. Remember that to express the ex-NOR gate function, the negation sign must be above the entire expression and within the brackets.

TIMING DIAGRAM

Figure 3-36 is a timing diagram for an ex-NOR gate.

Digital Circuits and Devices

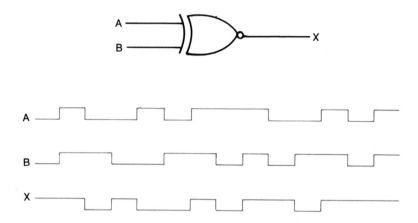

Figure 3-36. Exclusive-NOR timing diagram

As can be seen from the timing diagram, the only time the output is *high* is when the two inputs are the same levels. When the input levels are different, the output will be *low*.

Topic Review   3-8

1. The ex-NOR gate will produce a logic _____ output when all its inputs are the same level.

2. The Boolean expression for the IC ex-NOR gate is _____ .

Answers:

1. *high*
2. $X = \overline{[A \oplus B]}$

3-9 Gate Inversion

Gate inversion has three different forms: *sign; sense* and *sign;* and *sense*. The purpose of the three types of gate inversion is to achieve any gate function from any other type of gate, just by using inverters on the inputs, output or a combination of both. Gate inversion is very useful when only one type of gate function is available, but another gate function is needed.

Figure 3-37 shows an easy method of remembering which type of gate inversion to use when going from one gate function to another.

3
Digital Gates

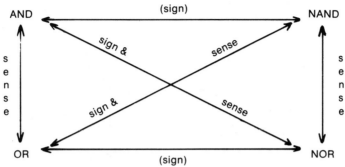

Figure 3-37. Gate inversion chart

SIGN INVERSION

The first and least complicated type of inversion is called sign inversion. Sign inversion allows you to get the inverse gate function when it is performed. **To perform sign inversion, you only invert the output of the gate.**

Sign inversion allows a NAND to operate as an AND, an AND as a NAND, an OR as a NOR, and a NOR as an OR.

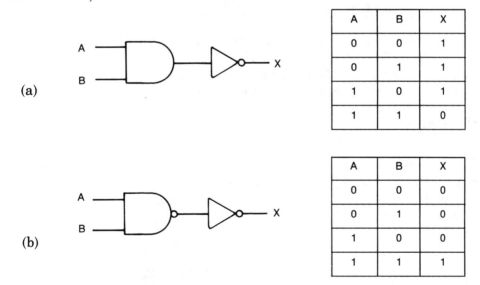

Figure 3-38. AND to NAND and NAND to AND gate inversion

As can be seen from the truth table in **Figure 3-38a**, the AND gate with an inverted output operates the same as a NAND gate. As can be seen from the truth table in **Figure 3-38b**, the NAND gate with an inverted output operates the same as an AND gate.

67

Digital Circuits and Devices

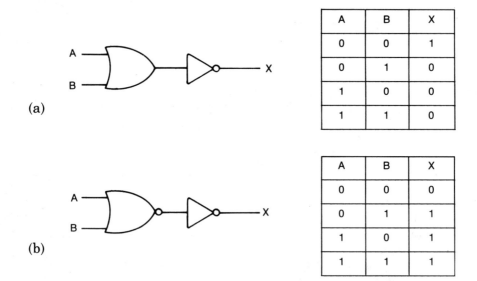

Figure 3-39. OR to NOR and NOR to OR gate inversion

The example shown in **Figure 3-39a** is an OR gate function converted into a NOR gate function, by inverting the output of the OR gate.

In the final example of sign inversion **Figure 3-39b** shows a NOR gate with an inverted output operating the same as an OR gate.

SENSE INVERSION

The second type of gate inversion is called sense inversion. Sense inversion allows you to get the opposite gate function, when it is performed. To perform sense inversion you must invert all inputs and also the output of the gate.

Sense inversion allows an AND to operate as an OR, an OR as an AND, a NOR as a NAND and a NAND as a NOR.

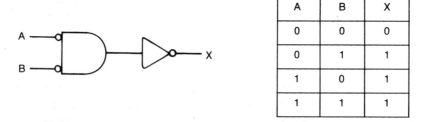

Figure 3-40. AND to OR gate inversion

As can be seen from the truth table in **Figure 3-40,** the AND gate with inverted inputs and output, operates the same as an OR gate.

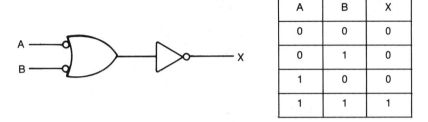

A	B	X
0	0	0
0	1	0
1	0	0
1	1	1

Figure 3-41. OR to AND gate inversion

As can be seen from the truth table in **Figure 3-41,** the OR gate with inverted inputs and output, operates the same as an AND gate.

In the third example, as shown in **Figure 3-42,** when the NAND gate has inverted inputs and output, the circuit operates the same as a NOR gate.

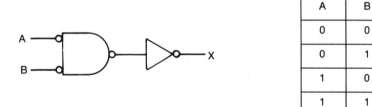

A	B	X
0	0	1
0	1	0
1	0	0
1	1	0

Figure 3-42. NAND to NOR gate inversion

In the last example of sense inversion (see Figure 3-43), when a NOR gate has inverted inputs and output, the circuit will operate the same as a NAND gate.

So to perform sense inversion, just invert all inputs and the output; this will yield the opposite gate function of the actual gate.

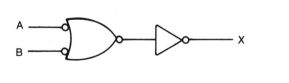

A	B	X
0	0	1
0	1	1
1	0	1
1	1	0

Figure 3-43. NOR to NAND gate inversion

Digital Circuits and Devices

SIGN & SENSE INVERSION

The final type of gate inversion is called sign & sense inversion; this is a combination of sign inversion and sense inversion. **To perform sign & sense inversion invert all inputs.**

Sign & sense inversion allows a NAND to operate as an OR, an OR as a NAND, a NOR as an AND and an AND as a NOR.

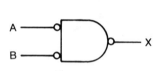

A	B	X
0	0	0
0	1	1
1	0	1
1	1	1

Figure 3-44. NAND to OR gate inversion

As can be seen from the truth table in **Figure 3-44,** the NAND gate with inverted inputs will operate the same as an OR gate.

Figure 3-45 illustrates the second example; an OR gate with inverted inputs will have the same truth table as a NAND gate.

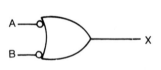

A	B	X
0	0	1
0	1	1
1	0	1
1	1	0

Figure 3-45. OR to NAND gate inversion

The third example is a NOR gate with inverted inputs and output; it will have the same truth table as an AND gate (see **Figure 3-46**).

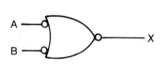

A	B	X
0	0	0
0	1	0
1	0	0
1	1	1

Figure 3-46. NOR to AND gate inversion

3
Digital Gates

As can be seen from **Figure 3-47,** an AND gate with inverted inputs will have the same truth table as a NOR gate.

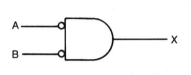

A	B	X
0	0	1
0	1	0
1	0	0
1	1	0

Figure 3-47. AND to NOR gate inversion

So to perform sign & sense inversion, just invert all inputs; this will yield the opposite inverted gate function of the original gate.

Topic Review   3-9

1. Gate inversion has three different forms: _____ ; _____ and _____ ; and _____ .

2. To perform sign inversion, you only invert the _____ of the gate.

3. Sign inversion allows a NAND to operate as an _____ , an AND as a _____ , an OR as a _____ and a _____ as an OR.

4. To perform _____ inversion you must invert all inputs and the output.

5. Sense inversion allows an AND to operate as an _____ , an OR as an _____ , a _____ as a NAND and a NAND as a _____ .

6. To perform sign and sense inversion invert all _____ .

7. Sign and sense inversion allows a NAND to operate as an _____ , an _____ as a NAND, a _____ as an AND and an AND as a _____ .

Answers:

1. sign, sign and sense, sense
2. output
3. AND, NAND, NOR, NOR

71

Digital Circuits and Devices

4. sense
5. OR, AND, NOR, NOR
6. inputs
7. OR, OR, NOR, NOR

3-10 Summary Points

1. The three basic gates that make up all digital circuitry are the AND, OR and NOT gate.

2. The only time the output of an AND gate goes *high* is when all its inputs are *high*.

3. The output of an OR gate will go *high* when any of its inputs go *high*.

4. The output of an inverter (NOT gate) will always be the opposite condition as its input.

5. The output of a NAND gate will be *high* when any of its inputs are *low*.

6. The output of a NOR gate will be *high* only when all its inputs are *low*.

7. The output of an ex-OR gate will be *high* only when its inputs are at different levels.

8. The output of an ex-NOR gate will be *high* only when its inputs are at the same level.

9. Gate inversion is used to achieve any gate function from any gate, by the use of inverting inputs, outputs or a combination of both.

10. A timing diagram will show the output condition of a circuit with different types of input levels.

3-11 Chapter Progress Evaluation

Match the following logic truth tables with their proper gate functions. Where A and B are inputs and X is the output.

3
Digital Gates

1.
A	B	X
0	0	0
0	1	1
1	0	1
1	1	0

= _____

2.
A	B	X
0	0	0
0	1	0
1	0	0
1	1	1

= _____

3.
A	B	X
0	0	1
0	1	0
1	0	0
1	1	0

= _____

4.
A	B	X
0	0	1
0	1	0
1	0	0
1	1	1

= _____

Digital Circuits and Devices

5. | A | B | X |
    |---|---|---|
    | 0 | 0 | 1 |
    | 0 | 1 | 1 |
    | 1 | 0 | 1 |
    | 1 | 1 | 0 |

    = _____

6. | A | B | X |
    |---|---|---|
    | 0 | 0 | 0 |
    | 0 | 1 | 1 |
    | 1 | 0 | 1 |
    | 1 | 1 | 1 |

    = _____

a) AND
b) OR
c) ex-NOR
d) NAND

e) NOR
f) ex-OR

Match the following timing diagrams with their proper gate function. Where A and B are inputs and X is the output.

7.

8.

a) AND
b) ex-NOR

Chapter 4

Logic Development

Objectives

Upon completion of this chapter, you should be able to do the following:

* Derive a Boolean expression from any gate circuit

* Derive a gate circuit from any Boolean expression

* Derive a truth table from any size gate circuit

* Derive a Boolean expression and therefore the gate circuit from any truth table

4-1 Introduction

One very important ability to have in digital electronics is the ability to derive a Boolean expression from a gate circuit. This coupled with the ability to perform Boolean algebra allows a technician to reduce if possible, the number of gates a circuit needs to perform a particular function.

Some companies use logic priority rules of operation for digital gates in deriving the Boolean expression for a gate circuit. Policies vary from company to company, so in this section we will derive Boolean expressions from gate circuits using a method that makes no assumption as to the order of logical operation. By making no assumptions when deriving Boolean expressions, each gate function must be placed within brackets which will represent the output function. When the circuit becomes larger, the outputs of gates are then used as inputs to other gates with their output function being enclosed in brackets. When a NOR, NAND or ex-NOR gate function is expressed, the negation sign must be placed inside the brackets. This allows no doubt that the expression represents one gate, rather than a gate with an inverter on its output. External inverters on the input or output of a gate is expressed with a negation line and does not require a set of brackets.

Digital Circuits and Devices

RULES

No assumption rules for deriving Boolean expressions from gate circuits are as follows:

1. The inputs of each gate will be enclosed in brackets along with the logic function or sign of the gate which represents the operation of the gate.
2. Gates with built-in inverted functions such as NOR, NAND and ex-NOR, must have their negation lines within the brackets.
3. External inverters on either the input or output of the gate are represented by a negation line over the input or output. External inverters do not require brackets for their designation.
4. Begin at the input of the circuit and work towards the output of the final gate in the circuit.

BOOLEAN EXPRESSION REVIEW

Before we begin deriving Boolean expressions from large multi-gate circuits, review the Boolean expression for the basic logic gates listed in **Table 4-1**.

Table 4-1.
Basic gates and their Boolean expressions

Gate Function	Logic Symbol	Boolean Expression
2-input AND gate		$X = [AB]$ or $X = [A \cdot B]$
2-input NAND gate		$X = [\overline{AB}]$
2-input OR gate		$X = [A + B]$
2-input NOR gate		$X = [\overline{A + B}]$
2-input ex-OR gate		$X = [A \oplus B]$
2-input ex-NOR gate		$X = [\overline{A \oplus B}]$
NOT or Inverter gate		$X = \overline{A}$

4
Logic Development

Application of Rules

Now let us derive the Boolean expressions of some multi-gate circuits applying the given rules and the information provided in **Table 4-1**.

As the first example, consider the multi-gate circuit of **Figure 4-1**. Notice that the B input is inverted by the inverter (circle) symbol on the input and then is ANDed to input A. The output of the circuit is equal to $A\overline{B}$ ANDed together. Because external inverters do not require brackets, there is no need to enclose B within brackets, just make sure the negation line covers the letter B.

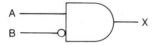

Figure 4-1. Boolean expression is $X = [A\overline{B}]$

From this example, notice that an inverter driving the input of a gate is represented as a circle at the input of the gate being driven. However, when the output of a basic gate requires external inversion, the full invertor symbol must be used.

For example, consider the multi-gate circuit of **Figure 4-2.**

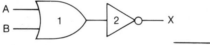

Figure 4-2. Boolean expression is $X = \overline{[A + B]}$

Note that the OR gate has an external inverter on its output, so first derive the Boolean expression for the OR gate and enclose it in brackets. Therefore, the output of the OR gate (identified as gate 1) is equal to $[A + B]$. The inverter (identified as gate 2) is driven by the output of the OR gate, so placing a negation line above the entire input Boolean expression to gate 2 adequately represents the Boolean expression at the X output of this multi-gate circuit. Since gate 2 is an inverter that is not built into the OR gate, no additional brackets are needed to complete the final Boolean expression for this multi-gate circuit.

From these two examples, the Boolean expression for the multi-gate circuit of **Figure 4-3** should be easy to derive.

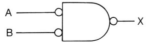

Figure 4-3. Boolean expression is $X = [\overline{\overline{A}\overline{B}}]$

In this example, both A and B inputs are inverted before they are NANDed together by the NAND gate. A negation line then should cover both input A and B,

77

Digital Circuits and Devices

and since each input is inverted separately, a negation line is needed for each input. Now the inputs for the NAND gate are in reality $\overline{A}$ and $\overline{B}$, which are then NANDed together. Therefore, the Boolean expression for the output of this multi-gate circuit is $X = [\overline{\overline{A}\overline{B}}]$.

For the multi-gate circuit of **Figure 4-4**, gate 1 is a two-input AND gate with input A inverted, yielding $\overline{A}B$ as the output. The output of this gate is only one input of the final gate. Gate 2 is a two-input AND gate with input B inverted, yielding $A\overline{B}$ as the output. This is the second input for the final gate. In the final gate (gate 3) the output of gates 1 and 2 are ORed together yielding $X = [[\overline{A}B] + [A\overline{B}]]$ as the Boolean expression for the output of this multi-gate circuit.

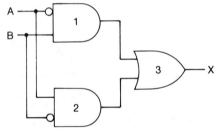

Figure 4-4. Boolean expression is $X = [[\overline{A}B] + [A\overline{B}]]$

An easy way of writing down the Boolean expression is to simply use the expression for each input, place the proper logic sign between each input, and then enclose the expression in brackets.

As a final example, consider the multi-gate circuit of **Figure 4-5** which consists of five basic gates.

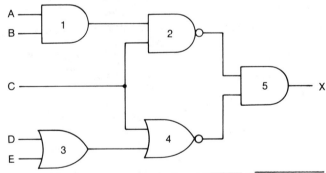

Figure 4-5. Boolean expression is $X = [[\overline{[AB]C}][\overline{[D + E] + C}]]$

The A and B inputs applied to gate 1 are ANDed together yielding [AB]. The

4
Logic Development

output gate 3 is [D + E]. The C input logic level is routed to both gates 2 and 4. The output of gate 2 is the C input NANDed with the output Boolean expression of gate 1, whereas the output of gate 4 is the C input NORed together with the output of gate 3. Finally, the outputs of gates and 4 are ANDed together with gate 5, which yields the final Boolean expression for this multi-gate circuit.

Topic Review   4-1

Derive the Boolean expression for each of the following multi-gate circuits.

1.

2.

3.

4.

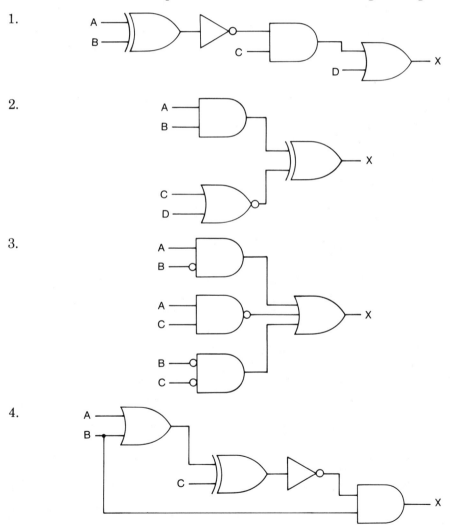

79

Digital Circuits and Devices

5.

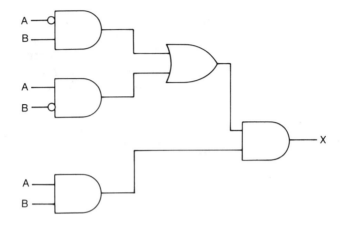

Answers:

1. X = [[$\overline{[A \oplus B]}$C] + D]
2. X = [[AB] $\oplus$ $\overline{[C + D]}$]
3. X = [[A$\overline{B}$] + [$\overline{A}$C] + [$\overline{B}\overline{C}$]]
4. X = [[$\overline{[A + B]} \oplus$ C] B]
5. X = [[[$\overline{A}$B] + [A$\overline{B}$]] [AB]]

4-2 Deriving Gate Circuits From Boolean Expressions

In this section the process of deriving gate circuits from Boolean expressions will be studied. As in section 4-1 we will not make assumptions as to how the Boolean expression was derived.

RULES

1. Start with the inner most brackets and work your way out towards the outer most brackets or until all inputs have been included in the gate circuit.
2. NORs, NANDs and ex-NORs will have a negation line inside the brackets.
3. If a negation sign goes beyond the brackets it is an external inverter.

Application of Rules

Now let's apply these rules to derive a multi-gate circuit from a given Boolean expression.

As our first example, consider the Boolean expression X = [[AB] C]. Notice that

4
Logic Development

X is equal to AB ANDed and then that output is ANDed to input C. The resulting circuit is shown in **Figure 4-6.**

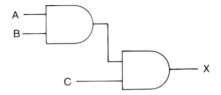

Figure 4-6. Multi-gate circuit for X = [[AB] C]

This multi-gate circuit is equivalent to a single three-input AND gate, where only the number of gates required for each circuit is different.

Now let's derive the multi-gate circuit for the Boolean expression X = [[A$\overline{\text{B}}$] + [$\overline{\text{C + D}}$]]. Starting with the innermost brackets notice that input B is inverted before it is ANDed to input A. This AND gate acts as one input to the final gate. The second input for the final gate is the output from a 2-input NOR gate with C and D as inputs. The final gate is a two input OR gate with the output of the 2-input AND gate as one input and the output of the 2-input NOR gate as the other input. The resulting multi-gate circuit is shown in **Figure 4-7.**

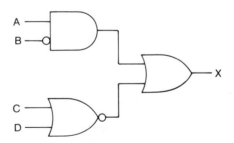

Figure 4-7. Multi-gate circuit for X = [[A$\overline{\text{B}}$] + [$\overline{\text{C + D}}$]]

As a final example, consider finding the multi-gate circuit for the Boolean expression X = [[[[A + B + E] [$\overline{\text{C + D}}$]] + A + $\overline{\text{B}}$] DE]. Going to the inner-most brackets, we find a three-input OR gate with A, B, and E as its inputs along with a NOR gate with C and D as its inputs. Notice that the output from both of these gates are ANDed together. The output from this two-input AND gate is ORed along with A and $\overline{\text{B}}$ as inputs. An inverter is required to obtain $\overline{\text{B}}$ from B. Finally, the output from the three-input OR gate is ANDed together with the D and E inputs. **Figure 4-8** shows the resulting multi-gate circuit.

81

Digital Circuits and Devices

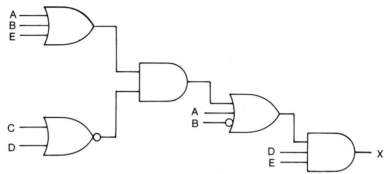

Figure 4-8. $X = [[\,[[A + B + E][\overline{C + D}]] + A + \overline{B}]\,DE\,]$

Topic Review 4-2

Derive the gate circuit for the following Boolean expressions.

1. $X = [\,[\overline{A} + B] + C + [AB]\,]$
2. $X = \overline{[\,[[ABC] + [ABC]]\,CB\,]}$
3. $X = [\,[\overline{[A + B]} + D]\,EF\,]$
4. $X = [\,[\,[\overline{AB}]\,[\overline{CD}]\,] + E + [AB]\,]$
5. $X = [\,[\,[\,[A + B + \overline{C}] + [AC]]\,AB] + \overline{C}]$

Answers:

1.

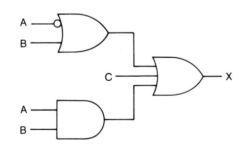

2.

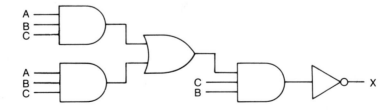

4
Logic Development

3.

4.

5.

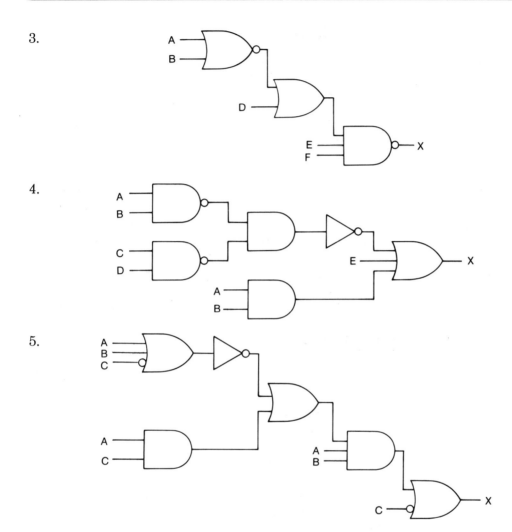

4-3 Deriving Complex Truth Tables

TRUTH TABLE REVIEW

When working with digital circuitry, technicians will often have to derive a truth table from a complex gate circuit. Normally, all that is required for the construction of a truth table is being able to find the output logic state for each possible combination of input logic states. When deriving a truth table, it is also very helpful to list the output for *each* gate so the derived condition for the next gate is easier to find. In the final truth table, only the last gate output is recorded.

Digital Circuits and Devices

APPLICATION OF TRUTH TABLE PRINCIPLES

As our example, let's find the truth table for the gate circuit of **Figure 4-9a**. The best way to tackle this problem is to break it into pieces by developing a preliminary truth table which shows the output logic state of each gate for all possible input combinations. This result is illustrated in **Figure 4-9b**. Notice that the starting point involves identifying all initial logic inputs and then writing down all possible input combinations. In this case there are only two inputs being A and B. From the logic state combinations of these two inputs, now determine the output logic states of each successive gate. Starting with gate 1, its output is [A$\overline{B}$]. Therefore, inverting the B input yields the $\overline{B}$ column of the preliminary truth table. ANDing the A and $\overline{B}$ columns yields the output logic states of gate 1. Next, notice that the output of gate 2 is [$\overline{A}$B], so inverting the A input yields the $\overline{A}$ column. ANDing the $\overline{A}$ and B columns yields the logic states of the $\overline{A}$B column. Finally, the output of gate 3 is X = [[A$\overline{B}$] + [$\overline{A}$B]]; therefore, the logic states of A$\overline{B}$ + $\overline{A}$B column are arrived at by ORing the A$\overline{B}$ and $\overline{A}$B columns. From this resulting preliminary truth table, the final truth table of **Figure 4-9c** is easily obtained.

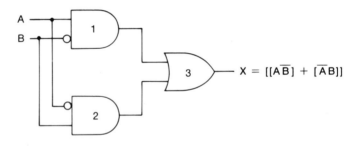

(a) Multi-gate circuit

Inputs		Gate Outputs				
A	B	$\overline{B}$	A$\overline{B}$	$\overline{A}$	$\overline{A}$B	A$\overline{B}$ + $\overline{A}$B
0	0	1	0	1	0	0
0	1	0	0	1	1	1
1	0	1	1	0	0	1
1	1	0	0	0	0	0

(b) Preliminary truth table

Figure 4-9. Deriving the truth table of a multi-gate circuit

4
Logic Development

Inputs		Final Gate Output
A	B	X
0	0	0
0	1	1
1	0	1
1	1	0

(c) Final truth table

Figure 4-9. (continued)

Topic Review   4-3

Find the preliminary and final truth tables for the following multi-gate circuits.

1.

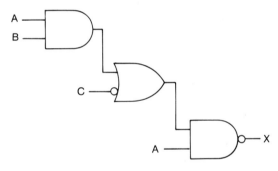

2.

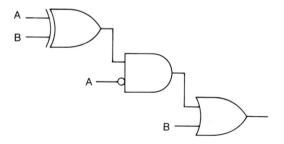

85

Digital Circuits and Devices

Answers:

1.

Inputs			Gate Outputs				Inputs			Final Output
A	B	C	AB	$\overline{C}$	AB + $\overline{C}$	$\overline{|AB + \overline{C}|}$|A|	A	B	C	X
0	0	0	0	1	1	1	0	0	0	1
0	0	1	0	0	0	1	0	0	1	1
0	1	0	0	1	1	1	0	1	0	1
0	1	1	0	0	0	1	0	1	1	1
1	0	0	0	1	1	0	1	0	0	0
1	0	1	0	0	0	1	1	0	1	1
1	1	0	1	1	1	0	1	1	0	0
1	1	1	1	0	1	0	1	1	1	0

2.

Inputs		Gate Output				Inputs		Final Output
A	B	A⊕B	$\overline{A}$	[A⊕B][$\overline{A}$]	[A⊕B][$\overline{A}$] + B	A	B	X
0	0	0	1	0	0	0	0	0
0	1	1	1	1	1	0	1	1
1	0	1	0	0	0	1	0	0
1	1	0	0	0	1	1	1	1

4-4 Deriving Gate Circuits From Truth Tables

Another important function a technician should be able to perform in digital electronics is to analyze a truth table and derive a gate circuit that will function the same as the original circuit. In some cases, the circuit will require less gates than the original circuit.

There are two methods for deriving a gate circuit from a truth table. The first method will use a *high* on the output to determine the circuit. The second method will use a *low* on the output to determine the circuit. Which method is used will be determined by which logic level (*high* or *low*) occurs more often in the truth table.

HIGH OUTPUT METHOD

The rules for the *high output method* are outlined as follows:

1. Each time the output goes *high* AND all inputs together. If the input is *high* use the input directly, if the input is *low* invert the input before it is ANDed,

86

4
Logic Development

with the other inputs.
2. OR each AND gate together, if there is more than one AND gate. Otherwise, just use the single AND gate.
3. Note all conditions of the truth table must be taken into account.
4. Use this method if the output of the truth table is *low* more than it is *high*.

Now let's apply these rules to the truth table of **Figure 4-10a.** Notice that there are two occurences when the X output is at a *high* logic state or level. In one case, the X output is a logic *high* when the A and B inputs are at logic 0 and 1, respectively. Therefore, X = logic 1 for [$\overline{A}$B]. The other occurence of a *high* logic output is when the A and B inputs are logic 1 and 1, respectively. Thus, X = logic 1 for [AB]. ORing these two expressions yields the Boolean expression X = [[$\overline{A}$B] + [AB]]. From this Boolean expression, the derived gate circuit of **Figure 4-10b** is obtained.

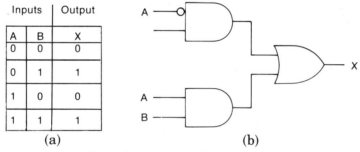

Inputs		Output
A	B	X
0	0	0
0	1	1
1	0	0
1	1	1

(a)   (b)

Figure 4-10. Derived gate circuit from truth table using the high output method

LOW OUTPUT METHOD

The rules for the *low output method* are outlined as follows:

1. Each time the output goes to *low*, AND the inputs together. If the input is *high* use the input directly, if the input is *low*, invert the input before it is ANDed with the other inputs.
2. OR all AND gates together, if there is more than one AND gate otherwise just use the single AND gate.
3. Invert the output of the Boolean expression.
4. All conditions in the truth table must be taken into account.
5. Use this method if the output is *high* more than it is *low*.

Now let's apply these rules to the truth table of **Figure 4-11a**. The two occurences when the X output is at a logic 0 is when the A and B inputs are at logic 1, 0 and 0, 0 respectively. Therefore, X = logic 0 for either [$\overline{A}$$\overline{B}$] or [A $\overline{B}$], which yields the Boolean expression X = $\overline{[[\overline{A}\overline{B}] + [A\overline{B}]]}$. From this Boolean expression, the derived gate circuit of **Figure 4-11b** is found.

87

Digital Circuits and Devices

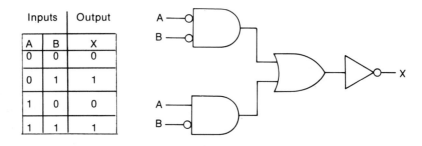

Figure 4-11. Derived gate circuit from truth table using the low output method.

Note that even though the derived gate circuits of **Figures 4-10b** and 4-11b are different, they perform the same logic function or operation since they were derived from the same truth table.

Topic Review   4-4

1. Derive the gate circuit from the given truth table using the *low* output method.

Inputs		Output
A	B	X
0	0	1
0	1	0
1	0	1
1	1	1

2. Derive the gate circuit from the given truth table using the *high* output method.

Inputs			Output
A	B	C	X
0	0	0	1
0	0	1	0
0	1	0	0
0	1	1	0
1	0	0	0
1	0	1	0
1	1	0	0
1	1	1	0

4

Logic Development

Answers:

1.

2.

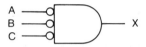

4-5 Summary Points

1. When deriving Boolean expressions from gate circuits, start at the input of the circuit and work your way towards the output of the circuit. Use the input of each gate along with the logic sign to determine the output of the gate. Enclose the output of each gate in brackets, unless the gate is an external inverter. If the gate has a built-in inverter function, the negation line must be within the brackets.

2. When deriving a gate circuit from a Boolean expression, start at the innermost brackets and work your way towards the outermost brackets until all inputs have been taken into account. Use the same rules for determining the type of gate as in section 4-1.

3. When determining a truth table for a multi-gate circuit, it is easier to list all outputs in the truth table. After the preliminary truth table is complete, make a final truth table using only the inputs of the circuit and the output of the circuit.

4. When deriving a gate circuit from a truth table, two methods can be used. In the *high* output method, AND the input conditions that will cause the output to go *high*. If the output goes *high* more than once, OR all the AND gates together. In the *low* output method, AND the input conditions that will cause the output to go *low*. If the output goes *low* more than once, OR all the AND gates together, and in either case invert the output of the circuit. Both of these methods assume positive logic.

4-6 Chapter Progress Evaluation

1. Derive the gate circuit from the following Boolean expression.

89

Digital Circuits and Devices

X = [[[A + B + C] [C + D]A] + E]

2. Derive the Boolean expression from the gate circuit.

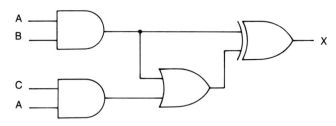

3. Derive the gate circuit from the given truth table below using the *low* output method.

Inputs			Output
A	B	C	X
0	0	0	1
0	0	1	1
0	1	0	1
0	1	1	1
1	0	0	1
1	0	1	1
1	1	0	0
1	1	1	0

4. Derive the Boolean expression for the above truth table using the *high* output method.

5. Derive the gate circuit from the Boolean expression in question 4.

6. Are the derived gate circuits of questions 5 and 3 functionally equivalent?

7. Derive the Boolean expression from the following gate circuit.

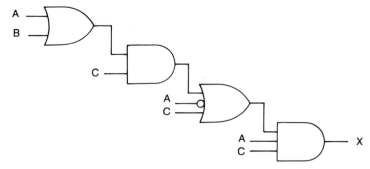

90

Chapter 5

Logic Families

Objectives

Upon completion of this chapter, you should be able to do the following:

* List the logic families within the saturated bipolar, nonsaturated bipolar, and the MOS logic groups

* Define terminology related to the specification of logic families

* Explain in detail the circuit operation of each TTL subfamily along with listing the advantages and disadvantages of each

* Describe the characteristics, advantages and disadvantages of the following logic families: RTL, DTL, HTL, TTL, CML, ECL, IIL, MOS and CMOS

* Describe interfacing techniques between logic families

* Describe handling techniques of logic families

* Determine functional operation of gates in negative logic

5-1 Introduction

Modern logic devices can be classified into three major groups:

1. Saturated Bipolar Logic
2. Nonsaturated Bipolar Logic
3. Metal-oxide-semiconductor MOS Logic

Within each of these major groups of logic devices, there are various logic families. A *logic family* is characterized mainly by the internal structure and composition of its logic devices. These logic families are tabulated in **Table 5-1**.

Digital Circuits and Devices

Table 5-1.
Logic Families

Saturated Bipolar Logic	Nonsaturated Bipolar Logic	MOS Logic
RTL	CML	NMOS
DTL	ECL	PMOS
		Low Threshold Logic
HTL	IIL or I^{2}L	VMOS
TTL		DMOS
		HMOS
		CMOS

Of the saturated bipolar logic group, resistor-transistor logic (RTL), diode-transistor logic (DTL), and transistor-transistor logic (TTL) have in common an operating voltage at or just below 5 VDC. They also present comparatively low impedance to the circuits they work in. High-threshold logic (HTL) resembles DTL in many respects, but requires somewhat higher operating voltages.

Of the nonsaturated bipolar logic group, ECL (emitter-coupled logic) and CML (current-mode logic) both work at low voltage levels—negative, in the case of ECL. Both also act with low circuit impedance. I^{2}L (integrated-injection logic) resembles none of the others, having special characteristics that suit large-scale integration.

Of the metal-oxide-semiconductor (MOS) logic group, all of its members need 12 to 15 VDC for operation. Furthermore, the logic families of this group exhibit high input and output impedance.

Some of these logic families can be broken down into subfamilies. For instance, the members of the TTL family are:

1. Standard TTL
2. High-speed TTL
3. Low-power TTL
4. Schottky TTL
5. Low-power Schottky TTL

High-speed TTL provides faster switching. Low-power TTL draws less supply current than is usual for standard TTL, and Schottky TTL incorporates clamping by special diodes.

Each of these logic families along with the subfamilies of TTL and CMOS will be presented in greater detail in later sections within this chapter.

5-2 Parameters of Logic Families

In order to understand how to choose a specific logic device from the various logic families, you must first understand the *parameters* used to describe the operation of logic gates. These parameters are outlined in the subtopics that follow.

Logic Families

VOLTAGES, CURRENTS, AND LOGIC LEVELS

The logic levels in any digital electronic system are dictated partly by DC supply voltages, and to some degree by the family of logic devices that dominate.

Table 5-2 shows the typical operating levels of logic subfamilies along with typical power dissipations. A lower power dissipation per gate is caused by each gate drawing less current from the DC power supply.

Table 5-2.
Typical operating levels of logic subfamilies

	RTL	DTL	TTL	HTL	NMOS	PMOS	CMOS, HMOS	ECL, CML**	I²L
Logic 0*	0 to 0.4	0 to 0.8	0 to 0.8	0 to 5	0 to 2	−5 to −15	2 to 12	−1.60	0.1
Logic 1*	2 to 5	2 to 5	2 to 5	10 to 15	5 to 12	0 to −2	1 to 4	−0.75	0.6
Operating supply voltage	+5	+5	+5	+15	+12	−15	+3 to +18	−5	+1
Power (mW) per gate	15	10	10	40	2	2	0.1	30	0.2

*Depends on circuit design. **Polarity of voltages depends on circuit design.

Notice that peak MOS logic levels are higher than TTL peak levels. Nevertheless, most modern MOS devices are TTL compatible. Therefore, logic in MOS circuits can maintain the same levels as in systems using all TTL. Families can therefore be mixed, yielding a wider choice of devices and greater design versatility.

ECL, CML, and I²L circuits work with extreme low logic levels — less than 0.8 V for ECL and CML, and only around 0.5 V for I²L. These small logic levels happen to be highly susceptible to noise.

SPEED OF DEVICE OPERATION

Bipolar and MOS transistors, like other semiconductor devices, face one serious operating limitation in digital applications: lack of speed. Various names have been applied to this characteristic, with a few of the more common being *switching time, operating speed,* and *propagation delay.* Other terms used are *switching rate, toggle rate, clock rate,* and *maximum operating frequency.*

No device can act instantaneously. Gate transition from one logic level to another may *seem* immediate, but some delay inevitably exists between stimulus and response. This time lag from input change (in logic) to output change is called *propagation delay.*

Digital Circuits and Devices

Propagation delay is measured from the point when the input-logic voltage reaches 50% of its change to the point when the output-logic voltage has attained 50% of its total change.

Figure 5-1 provides an illustration of propagation delay, which is usually written as t_{pd} or t_d in the specification sheets of manufacturers.

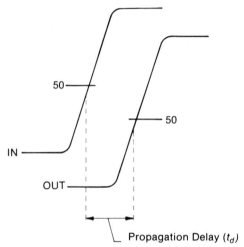

Figure 5-1. Propagation delay of a logic gate

Typical delays in CMOS gates range from 20 to 50 *nsec* (nanoseconds), depending upon internal construction. I^2L gates are very slow, in most instances, worse than CMOS. DTL devices show about the same propagation delay as CMOS, with RTL a bit faster.

TTL works much faster. Typical delay for TTL is 10 to 20 *nsec*. However, certain types of TTL such as Schottky TTL are as fast as 3 to 5 *nsec*. Low-power versions of TTL are the slowest types of TTL, with delays of 33 *nsec*. Low-power Schottky gates switch at about the same speed as standard TTL, but with much reduced power consumption.

Of all logic types, ECL switches the fastest. These devices operate on a linear portion of their transfer curve, thus avoiding the saturation that slows down other types of logic gates. Logic can therefore change in incredibly short times, with the propagation delay in some ECL gates being less than 1 *nsec*.

Translating propagation delays into maximum switching rate depends to some extent on where the input comes from and what the output is fed to. As a general rule of thumb, you can divide 350 by the amount of propagation delay (in nanoseconds) and arrive at a rough estimate of the maximum clock rate or operating frequency in megahertz (*MHz*).

According to this calculation, CMOS and I^2L devices can be driven by sequential logic signals that switch at 5 to 12 *MHz*. Ordinary TTL works fine at 30 to 40 *MHz*, and special TTL devices operate at 50 *MHz* and better. ECL can be depended on beyond 300 *MHz*, some devices as high as 1000 *MHz*.

The frequency limit of a digital system will depend on the propagation delays of the various gates making up the system. Also, the propagation delays of logic gates tend to be additive within a digital system.

NOISE IMMUNITY

Since logic levels are low voltage levels to begin with and since they switch very fast, electronic noise can pose a real problem for gate devices, both bipolar and MOS. *Noise* can be defined as any unwanted, unintentional voltage change. The amplitude and the time duration of a noise pulse depend upon its source.

Much noise that affects digital circuitry originates externally, and is carried into the equipment by power supply lines. Some types of noise, as from static discharges, reach logic circuits by radiation.

Inside the gate itself, the sudden transition of switching from *low* to *high* may create noise. Changes in loading on DC supply lines, whether abrupt or slow, can introduce noise to all the stages and circuits fed by those lines.

A gate may be said to exhibit excellent *noise immunity* when its logic state changes only in accord with proper logic inputs and not in response to random signals. This ability of a logic gate to reject noise can be expressed by a specific voltage value, usually known as the *noise margin*.

In a TTL device, for example, logic 0 is ordinarily defined as any voltage level between 0 *V* and 0.6 *V*. Similarly, most definitions place logic 1 between 2.2 *V* and 4.9 *V*. These logic ranges apply at either input or output and also assume positive logic.

As you can see, a region of *undefined* operation lies between logic 0 and logic 1. A noise voltage added to logic 0 input, if large enough, could push the gate into this area of uncertain operation. As a result, its output level might switch, as if the device were seeing logic 1 input. The gate might even try to jump back and forth between the two states. Noise added to logic 1 input could have similar undesirable results.

The duration of a noise pulse may be only nanoseconds. Yet if it exceeds the switching time for the gate, an unintended change of state could be passed along to the output.

In a practical logic device, the manufacturer controls logic thresholds through the twin techniques of material doping and junction structure. Thus, for a particular gate, the designer might specify an *input upper threshold* for logic 0 of 0.8 *V*. If so, a noise spike of 0.2 *V* peak could accompany a logic 0 voltage of 0.6 *V* without pushing operation into the undefined region. This 0.2 *V* would be the *low noise margin* for the device because it applies to the low-logic condition.

In the same device, an *input lower threshold* for logic 1 might be specified as 1.8

Digital Circuits and Devices

V. A negative-going noise pulse as high as 0.4 V peak would still not force a logic 1 signal at 2.2 V low enough to be outside the proper logic 1 range. This extra 0.4 V constitutes the *high noise margin*, since it affects logic-high states.

If only one noise margin is quoted in a spec sheet, it's always the lower of the low and high noise margins.

Typical TTL noise margin lies between 0.1 V and 0.3 V. HTL devices may offer a noise margin as great as 6.0 V; therefore, HTL gates are advantageous in noisy electrical environments. CMOS noise margins depend on the supply voltage used which then effects the logic level. For example, when operated with TTL the supply voltage will be 5 VDC, and a CMOS device may have a noise margin as good as 1.0 V. In addition, CMOS devices can deliver excellent noise immunity, since logic *high* can be placed as *high* as 15 V. ECL devices poorly exhibit very low noise margins, since their logic levels are extra low.

Actual noise performance can be improved by applications designers. Circuit arrangements should keep logic 1 well above its lower threshold and logic 0 well below its upper threshold.

OPERATING TEMPERATURES

In some ways, temperature can be one of the more critical logic device specifications. Junction devices generate considerable internal heat of their own, which must be dissipated through the packaging and case to surrounding air or other material. If such heat isn't dissipated, junction characteristics change, rendering operation unpredictable. Junctions also change character when subjected to excessive cold.

Field-effect devices, such as MOS gates, create less heat of their own, and their operation is therefore less affected by temperature. Still, heat developed from current in the channel must be dissipated through the package. MOS devices also operate poorly in cold temperatures since channel-carrier activity becomes sluggish.

Ceramic cases do a far more efficient job than plastic in getting rid of unwanted heat. Terminals themselves can also carry heat outside, where the air can dissipate it, so that a flatpack housing keeps internal devices cooler than a DIP case.

As a result of all this, you'll find temperature specifications common in catalogs of logic devices. A more or less ordinary operating temperature spec is $-30°C$ to $+75°C$. This corresponds to $-22°F$ to $+167°F$.

A more stringent temperature specification for logic devices stretches from $-55°C$ to $+125°C$ ($-67°F$ to $+257°F$). Devices rated for this expanded range are more expensive. Military users of logic ICs usually specify this wider range.

You'll also encounter some highly specific *typical* operating ranges. For example, devices intended for automotive environments usually specify a temperature range of $-40°C$ to $+85°C$. An industrial temperature specification usually ranges

Logic Families

from 0°C to +70°C since the temperature range within the industrial environment is not as extreme.

LOADING, FAN-IN, AND FAN-OUT

The load factors of a logic device are of critical importance when fast logic transitions are involved. Excessive load can distort the timing of logic sequences. The loading of both the input and output of the logic gate is normally considered. Input loading is usually the result of the combined capacitance of the input junction or gate, plus any stray lead capacitance. Input loading ultimately limits the ability of a device to accept (or reproduce) fast-rising (or fast-falling) logic transistions. Input capacitance, when listed, is typically on the order of a few picofarads (*pF*), and it is usually smaller for MOS devices than for TTL.

A specification called *loading factor* is listed by some manufacturers. **An input-loading factor designates how much load is presented to the preceding, or driving, device.** This loading factor may be less than 3 for MOS devices. Most TTL devices exhibit input-loading factors between 4 and 10.

Another input term is *fan-in*. Fan-in has an inverse relation to loading factor and refers to the number of devices that can be connected to the gate input. Generally, fan-in is 1. A fan-in spec of less than 1 means that the device presents less load than is normal for its type. Fan-in of 2 or more means that the device loads down the driver more than normally.

Output-load ratings spell out the ability of a device to drive other devices. A particular CMOS device might be rated as able to drive three ordinary TTL loads or two low-power Schottky TTL loads. The same device might be able to drive five or six CMOS devices. TTL devices seldom can handle more than one or two TTL loads.

The term *fan-out* best describes the output-drive capabilities. The CMOS device just described, for example, would be rated for a TTL fan-out of 3 (or a CMOS fan-out of 5). A TTL device that can handle only one load is said to have a TTL fan-out rating of 1.

Buffers may be inserted after either TTL or CMOS devices, to increase fan-out capability. Each buffer or driver adds its own fan-out capability to that of a driving gate which is usually a *spreading factor* of 3 to 5.

Many manufacturers use *loading factor* to express output capabilities. This specification relates numerically to the input-loading factor mentioned earlier. A CMOS device with an output-loading factor of 10, for example, can drive four devices that exhibit input-loading factors, respectively, of 2, 2, 3, and 3. For purposes of estimating, you can figure on input-loading factors of 4 for regular TTL, 6 for Schottky TTL, and 2 for most CMOS. Remember that these numbers are only approximate. Always study the spec sheets for the specific gates to avoid exceeding normal fan-out for a

Digital Circuits and Devices

device. Specification sheets for logic families will be discussed as the following topic in this chapter.

ECL devices have fewer loading problems than other families of logic. Being essentially current-operated rather than voltage-sensitive, ECL gates allow fan-outs of 20 or 30, or even higher. Ordinarily, you can't mix ECL with TTL and CMOS in a system without special translator circuitry to revise the logic levels. However, TTL-compatible ECL and I^{2}L devices that contain *logic translators* right on the chip are available. You must observe spec sheets carefully if you plan to use these mixed device types.

Topic Review   5-2

1. Various _____ are used to describe the operation of logic gates.

2. The _____ _____ in any digital electronic system are dictated by DC supply voltages and to some degree by the family of logic devices that dominates.

3. The time lag from an input change in logic level to output logic change is called _____ _____.

4. The ability of a logic gate to reject noise can be expressed by a specific voltage value known as _____ _____.

5. The term _____-_____ best describes the output drive capabilities of a logic gate.

Answers:

1. parameters
2. logic levels
3. propagation delay
4. noise margin
5. fan-out

5-3 Specification Sheets for Logic Families
MANUFACTURERS AND SPECS

A recent catalog from an IC manufacturer lists 86 distinct *varieties* of integrated circuits, 70 of them digital. Each of those 70 categories represents hundreds of individual devices, each with its own purpose. And that's just one of the several dozen

Logic Families

manufacturers that supply integrated circuits. So, with so many ICs to choose from, how can you possibly pick the right one for a new application or to improve an older piece of equipment?

You study the specifications, that's how. In the manufacturers' catalogs, you'll find specs presented in a number of different ways.

A *blanket* spec chart **Figure 5-2**, for example, covers many devices. Such charts usually show the highest allowed voltages and currents, and recommended (or typical) operating ranges. More complete charts name and describe each device, and may even contain additional specifications peculiar to individual devices.

MAXIMUM RATINGS (Voltages referenced to V_{SS})			
Rating	Symbol	Value	Unit
DC Supply Voltage	V_{DD}	−0.5 to +18	Vdc
Input Voltage, All Inputs	V_{in}	−0.5 to V_{DD} +0.5	Vdc
DC Current Drain per Pin	I	10	mAdc
Operating Temperature Range — AL Device CL/CP Device	T_A	−55 to +125 −40 to +85	°C
Storage Temperature Range	T_{stg}	−65 to +150	°C
RECOMMENDED OPERATING RANGE			
DC Supply Voltage	V_{DD}	+3.0 to +15	Vdc

Figure 5-2. Typical blanket specification chart

Another format presents specifications in essay form. The sheet pictured in **Figure 5-3**, for example, has been torn from a manufacturer's catalog. Specs are defined in detail, sometimes with comments on how they were derived.

More detail yet can be derived from an individual specification sheet. Usually, IC makers provide these only for complex designs, or for a unique IC dedicated to a special purpose.

Operating parameters, limits, and conditions are shown, often with notes describing typical applications. A schematic—or at least a block diagram—is also usually included, showing inside wiring.

If you can't find the data you need in a catalog or spec sheet, turn to the nearest electronic distributor (a few will help) or manufacturer's representative (*most* will help). From them, you may learn of the existence of a spec sheet or catalog newer or more detailed than what you have. If a manufacturer or his sales people can't or won't help, seek out a *second source* of the chip you wish to use. Second-source manufacturers sometimes do a better (and more willing) job of documenting products than the original developer of a device.

Digital Circuits and Devices

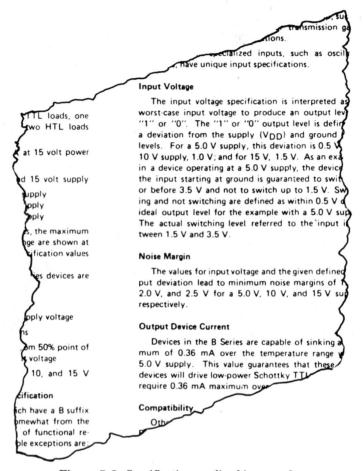

Figure 5-3. Specifications outlined in essay form

CROSS IDENTIFICATION FOR LOGIC ICS

Some technical books contain specifications of a general nature. They can be valuable aids. Let your local electronic distributor know you're interested in any literature he can provide. You can build a versatile library of at least minimal data.

One help in using such books capitalizes on numbering conventions that have evolved among logic ICs.

One of the oldest TTL families carries 7400-series numbers as identification; variants may incorporate letters as part of the number. A 7402 IC, for instance, contains four two-input NOR gates. A 74H02 contains the same gates, but in a higher speed, heavy-current construction. A 74S02 uses a kind of diode clamping called Schottky to improve performance. A 74LS02 is a low-current Schottky type. Finally,

Logic Families

a 74C02 IC offers the same gate configurations, but with CMOS technology.

This cross identification prevails among the chips from most manufacturers. A Motorola SN54LS03, for example, carries the same internal configuration, including pinout connections, as the 7403. The "LS" means it's low-power Schottky. The "54" stamps it as one of a Motorola series that can face extreme temperatures: −55°C to +125°C (the military range). The "03," finally, brands it as equivalent to the 7403.

This agreement in numbering carries to IC families of almost every type.

WHAT SPECS ARE IMPORTANT

Generally, you'll need most if not all of the following *key* information about a logic-type IC:

>Family, plus its compatibility with other families
>Operating voltages and power dissipation
>Logic levels (usually consistent within a family)
>Operating speed, or cutoff frequency
>Noise margin, or noise immunity
>Temperature range (if environment is unusual)
>Contents of the IC
>Pinout diagram
>Fan-in and fan-out (input-output loading)
>Case shape and composition (sometimes important)

Manufacturers have all this information in one form or another. Some supply it to any customer without question. Others don't, so you may have to seek the help of a distributor or rep. If you have a constant problem obtaining data, do your best to avoid that brand of IC; the manufacturer isn't delivering the kind of support to which you are entitled.

PINOUT DIAGRAMS

The best IC spec sheets and catalogs include *pinout diagrams* that show how the gates or flip-flops inside connect to terminal pins. Unfortunately, few manufacturers include this information anymore. Pinout diagrams seem to be available only for unusual ICs. So you'll probably need to keep your own general pinout diagram book: the kind you can buy from a book publisher.

Figure 5-4 illustrates the pinout diagram for a *hex inverter*. The word *hex* indicates there are six gates in the IC. In the TTL family, identification for this IC is 7404 and 74C04 in CMOS.

Digital Circuits and Devices

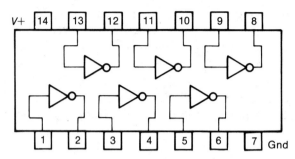

Figure 5-4. Pinout diagram for a hex inverter

Topic Review 5-3

1. A _____ _____ chart covers the specifications of many devices.

2. All of the parameters of logic families are important _____ of logic devices.

3. The best IC spec sheets and catalogs include _____ _____.

Answers:

1. blanket spec
2. specifications
3. pinout diagrams

5-4 Saturated Bipolar Logic Families

RESISTOR-TRANSISTOR LOGIC (RTL)

One effect of transistor buffering is that a logic state retains its original level; if the input logic 1 is 4.8 V, so is the output logic 1. Similarly, if input logic 0 is 0.4 V, output logic 0 takes the same value. Logic levels thus hold, despite a modest amount of loading on the output. Even if fan-out involves several other gates, logic levels stay the same.

Two simple RTL devices appear in **Figure 5-5**. The letters RTL stand for *resistor-transistor logic*. The devices represent saturated bipolar logic with R_1 and R_2 resistors added for isolation and to reduce current, which lowers the power consumption of the gate. The resistors in series with the OR gate inputs in **Figure 5-5a** limit base current and result in less fan-in loading. Output resistor R_3 may be made fairly high in value, keeping current drain low for the stage.

The gate at **5-5b** inverts the logic value normally resulting from inputs A and B, therefore performing the NOR function.

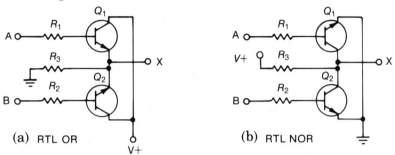

Figure 5-5. Examples of RTL gates

DIODE-TRANSISTOR LOGIC (DTL)

Diodes can't take much loading. This is why diode logic seldom appears alone. If diodes are part of a gate IC, resistors and at least one transistor are added.

The DTL schematic in **figure 5-6** might fool you at first glance since the three inputs are connected together. However, the diodes face the wrong way for an OR function. So then what logic does the circuit perform?

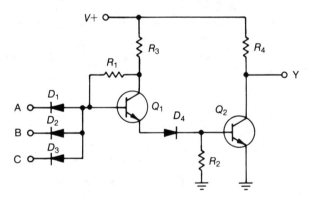

Figure 5-6. Example of DTL gate

If any input sees logic 0, its diode conducts and pulls bias on Q_1 below the threshold level of transistor conduction. But when A and B and C are all reverse-biased (logic 1), they have no effect. Positive bias applied through the large value of R_1 makes Q_1 conduct. Logic 1, from across R_2, appears at the base of Q_2. With Q_2 conducting, the voltage at its collector drops to near zero (Q_2 saturated). Q_2 in satura-

Digital Circuits and Devices

tion causes R_4 to develop approximately $V+$ across it. This results in a logic 0 at output Y. Therefore the output Y is at logic 0 only when all three inputs are at logic 1. Consequently, the DTL gate of **Figure 5-6** must be a NAND gate.

The DTL circuitry in a logic IC offers a significant improvement in noise immunity over RTL. Noise problems are almost always worse around logic 0—the low end of the input-voltage range. Diodes in series with the inputs mean that—even when A, B, and C are logic low—the base of Q_1 can't fall lower than 0.6 V, because of the junction drop in each diode. As a result, any noise spike arriving on the input lines must exceed this diode-junction drop; otherwise, it won't affect the logic stage at all. D_4 has a similar noise-isolating effect in series with the input of Q_2. Note that two diodes connected in series will raise the input threshold to 1.2 V which provides even better noise immunity.

HIGH-THRESHOLD LOGIC (HTL)

A need for even greater noise immunity in certain environments led to an extension of the series-diode isolation effect. When a diode is reverse-biased with a voltage high enough, its junction breaks down and allows current to flow (called *avalanche current*). However, backward resistance of the diode is high. A junction voltage develops, just as it does during forward operation of the diode. In a silicon junction, this *zener* voltage is very stable—always close to 6.8 V. A diode operated this way is called a zener diode. **Figure 5-7** shows a two-input NAND gate which uses high-threshold logic. The zener input established by D_3 places the turn-on voltage point of the transistor much higher than normal—hence, the term *high-threshold*. Incoming logic voltage must exceed the zener rating before it can reach the transistor base. With logic 1 so much higher than logic 0, a noise margin as high as 6 or 7 V is possible. This offers excellent noise immunity.

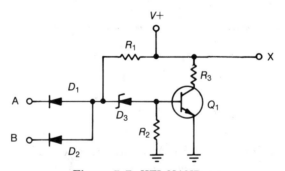

Figure 5-7. HTL NAND gate

Notice that high-threshold logic seems closer to DTL than to any other logic family within the saturated bipolar group.

5
Logic Families

TRANSISTOR-TRANSISTOR LOGIC (TTL)

TTL is the most popular logic family of the saturated bipolar logic group. Within the TTL family are various subfamilies of logic. These subfamilies of TTL will be discussed in greater detail in the following topic of this chapter.

Topic Review 5-4

1. The families comprising the saturated bipolar logic group are _____, _____, _____, and _____.

2. The saturated bipolar logic family which provides excellent noise immunity is _____.

3. The most popular logic family of the saturated bipolar logic group is _____.

Answers:

1. RTL, DTL, HTL, TTL
2. HTL
3. TTL

5-5 TTL Subfamilies

STANDARD TTL

Transistor-transistor logic strongly resembles DTL devices. In place of each input diode, a modern TTL device substitutes the emitter-base junction with a special NPN transistor.

Shown in **Figure 5-8** is the circuit for a four-input TTL NAND gate. Four N+ emitters have been diffused into the base of the input transistor, which is connected as a common-base amplifier.

Operation of the four-input TTL NAND gate is as follows. A logic 0 applied to any emitter causes base current to flow which results in the saturation of Q_1. With the collector of Q_1 going to logic 0, conduction in Q_2 is cut off which leaves Q_4 also turned off. In other words, a logic 0 applied to any input of the TTL NAND gate causes Q_1 to saturate which results in the cutoff of Q_2 and Q_4 (no collector current). Notice that when Q_2 and Q_4 are turned off, a load on the X output of the TTL gate will cause Q_3 to saturate leaving the X output at logic 1.

Only when all inputs see logic 1 does Q_1 turn off, which results in the conduction of Q_2 since it receives its base current path through the forward-bias of the base-to-collector junction of Q_1.

Digital Circuits and Devices

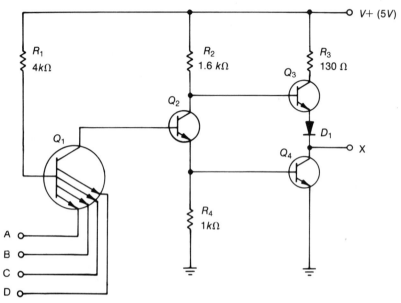

Figure 5-8. TTL NAND gate

The saturation of Q_2 causes the collector-to-emitter voltage of Q_2 to be slightly above ground which turns off Q_3. However, since Q_2 is operating in saturation, the collector voltage of Q_2 will be at logic 1 which turns on Q_4. The saturation of Q_4 results in the collector-to-emitter voltage of Q_4 to be slightly above ground which causes the X output to be pulled low to logic 0.

To summarize the operation of the four-input TTL NAND gate, a binary input pattern of 1111 results in a logic 0 output, whereas a binary input pattern consisting of one or more logic 0 inputs will produce a logic 1 output.

Notice the output arrangement of the TTL NAND gate. The two transistors in series across the DC supply is called a *totem-pole* output. Its main feature is transistor Q_3, which is called an *active pull-up* transistor that conducts only when at least one input is at logic 0. Since input loading of any logic gate is largely capacitive, the gate output must charge the capacitance of whatever load follows. Therefore, using an active pull-up transistor at the output of the TTL gate makes its output charging rate much quicker than any ordinary pull-up resistor could.

In addition, the pull-up transistor also makes the transition from logic 1 to logic 0 faster. Thus the totem-pole output presents less output impedance than ordinary output designs. This is one reason why TTL is so much faster than RTL or DTL.

HIGH-SPEED TTL

Fast as it is, the normal TTL rate isn't enough for some applications. Two steps can be taken to improve speed in one subfamily of TTL designs:

1. *Placing diodes at the inputs,* as in **Figure 5-9.** This damps out any negative-direction ringing caused by capacitive effects along the input circuits to the device. Logic transitions from logic 1 to logic 0 reach the emitters *clean.* In other words, the clamp diodes at the inputs improve noise immunity of the logic gate.
2. *Lowering output impedance even further by modifying the totem-pole output.* This is done by adding Q_5, a low-impedance driver for the active pull-up transistor; by eliminating D_1 of **Figure 5-8**, and by reducing R_3 even further.

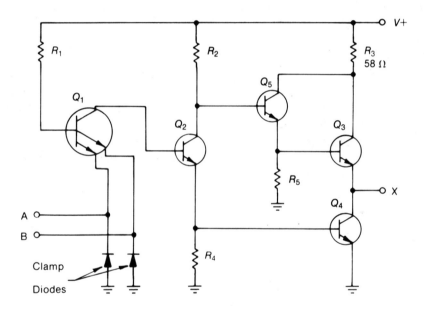

Figure 5-9. High-speed TTL NAND gate

Applying these two steps to standard TTL can double the switching rate of a TTL device since shortening the propagation delay of a device will contribute to faster operation. In addition, a tradeoff exists since high-speed TTL will dissipate more power than standard TTL.

LOW-POWER TTL

A typical way to reduce power consumption in a TTL device is by raising the resistor values. In particular, higher values of R_1 and R_2 reduce current drawn by the circuits.

However, doing so wipes out many of the advantages that lead to fast switching. In general, a trade-off can be made between high speed and low power consumption. But low-power TTL devices seldom permit switching rates in excess of 8 or 10 *MHz*.

SCHOTTKY TTL

Saturation logic itself limits switching speed. During saturation, current carriers inside a transistor tend to bunch up at the base-collector junction. When the input bias drops suddenly to zero, these excess carriers must drain off, producing a collector current for a few nanoseconds after base-emitter current has quit. This *junction charge* is largely responsible for the characteristic already described as *propagation delay*.

A special kind of diode, known as a *Schottky barrier diode,* has a junction structure that doesn't allow this concentration of carriers. Schottky diodes can be diffused into a monolithic substrate, right along with bipolar transistors. Placed in parallel with the base-collector junction of each transistor, a Schottky diode prevents this junction saturation. The result is a much shorter switching delay.

Figure 5-10a illustrates the wiring of a Schottky clamping diode across the *b-c* junction of an ordinary transistor. A diagram symbol for the transistor formed in this manner appears at **5-10b**.

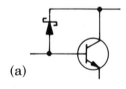
(a)

(b)

Figure 5-10. Using the Schottky barrier diode

TTL devices, such as the Schottky TTL NAND gate of **Figure 5-11**, are capable of switching rates well up toward 100 *MHz*. True, these diodes raise the power needs of a gate, but not as seriously as the other methods of increasing operating speed.

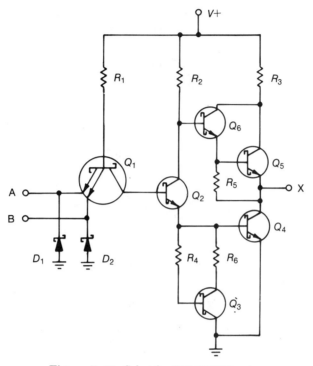

Figure 5-11. Schottky TTL NAND gate

LOW-POWER SCHOTTKY TTL

This subfamily of TTL devices has grown extremely popular since it combines low cost, low-power dissipation, and a high switching rate.

Lower-power Schottky TTL is only slightly faster than standard TTL, and requires less energy consumption. This is accomplished by using the Schottky clamping diodes along with raising the circuit impedance levels above those of regular Schottky TTL. However, this slows the switching operation down from that of regular Schottky TTL.

In more specific terms, the frequency limits of low-power Schottky TTL are 50 to 60 *MHz* which is less than regular Schottky TTL, but slightly better than standard or low-power TTL. On the other hand, the power needs of a low-power Schottky device are nearly double than that of a low-power TTL, yet it draws only about 20% as much current as ordinary TTL. For many applications, this trade-off is highly beneficial.

SUMMARY OF THE TTL SUBFAMILIES

In **Table 5-3,** the fan-out characteristics of TTL gates are listed. The sub-

Digital Circuits and Devices

families of TTL gates are shown under the *driver* heading which designates the output driving device. The *gates/loads* heading shows how many input gates each driver can effectively drive.

Table 5-3.
Fan-out characteristics of TTL gates

Driver	Gates/Loads			
	(—)	(S)	(L)	(LS)
Standard TTL (—)	10	8	80	40
Schottky TTL (S)	12	10	100	50
Low-power TTL (L)	2	1	20	9
Low-power Schottky TTL (LS)	5	4	40	20

Table 5-4 shows a comparison of the speed and power of each TTL subfamily relative to standard TTL.

Table 5-4.
A speed/power comparison of TTL subfamilies

TTL Subfamily	Speed	Power
Standard	x1	x1
High-speed	x2	x2
Low-power	x 1/10	x 1/10
Schottky	x 3.5	x2
Low-power Schottky	x1	x 1/5

More specifically, **Table 5-5** lists typical gate propagation delay and gate power dissipation of the TTL subfamilies.

Table 5-5.
Typical gate characteristics of TTL subfamilies

TTL Subfamily	Gate Propagation Delay in nanoseconds	Power Dissipation per gate in milliwatts	Max. Operating frequency in *MHz*
Standard	10	10	35
High-speed	6	22	50
Low-power	33	1	3
Schottky	3	19	125
Low-power Schottky	10	2	45

OPEN COLLECTOR TTL

Designers who put TTL logic ICs together in system circuitry observe certain precautions. Ordinarily, TTL logic outputs *cannot* be tied to a common point—unless their output states will always be alike at the same time, either *high* or *low*. However, an *open collector* is one kind of TTL IC structure that allows this "wired OR" connection. As shown in **Figure 5-12**, the open collector TTL NAND gate requires an external pull-up resistor to the DC supply voltage.

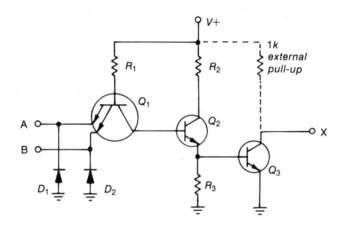

Figure 5-12. Open collector TTL NAND gate

A designer can wire together the outputs of two or more open collector gates without the concern over logic conflicts, provided this wiring is done as shown in **Figure 5-13**. Any open-collector gate needs a *pull-up resistor between its output and the operating voltage supply to establish a logic high level*. In *wired logic* (also called *tied logic*), this pull-up resistor must be placed between the gate's output pin and the tie point. A pull-up resistor (R_C) common to all outputs then connects the tie point to a DC supply. Values chosen for the resistors depend on current drawn by each gate and by all gates cumulatively. The values compromise between current during logic-*high* output and that during logic-*low* output.

Logic at X in **Figure 5-13**, depends on active conditions of the respective gates. You can construct a truth table for the entire stage to help you envision stage operation. **Remember that logic low at the output of any gate pulls logic low at the tie point.** Thus, X is *low* if A is *high*. Or, X stays *low* whenever B or C is *low*, since the AND gate can't output a *high* if either input drops *low*. Or, X goes *low* if D and E both go *high*.

Digital Circuits and Devices

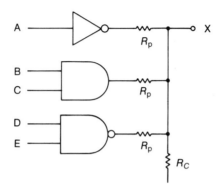

Figure 5-13. Wiring connection of open collector gates

In actual systems, most wired-logic stages contain only one kind of gate function. That is, all the devices are OR, NOR, AND, NAND, or NOT. This, of course, simplifies the truth table for a wired-logic stage.

TRI-STATE TTL SYSTEMS

Outputs that can be *three-stated* can be tied together, under one condition: the gates can't be allowed to go active simultaneously. If that were permitted, logic levels might again conflict.

Figure 5-14 demonstrates one way to prevent logic level conflict. Only one of the drivers can be active at a time. Logic 1 on the *enable* line turns on driver A, but three-stages (turns off) driver B—because of the inverting gate in that branch of the enable line to driver B. Logic 0 enable does just the opposite, activating driver B and inhibiting A. Such arrangements take advantage of time differences in clock pulses.

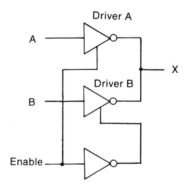

Figure 5-14. Application of tri-state devices

5
Logic Families

Figure 5-15 shows the internal circuitry of a tri-state TTL NAND gate and its corresponding logic symbol.

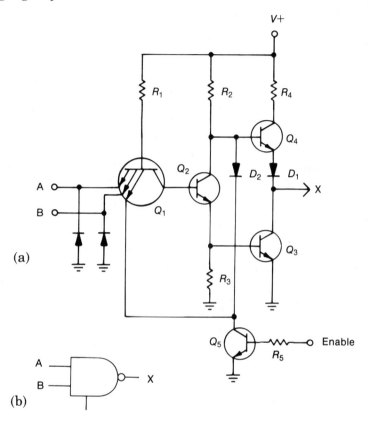

Figure 5-15. Tri-state TTL NAND gate

Topic Review 5-5

1. The subfamilies of TTL are _____, _____ - _____, _____ - _____, _____ and _____ - _____ _____.

2. High-speed TTL is accomplished by placing _____ _____ at the inputs and by lowering the _____ _____ of the totem-pole output.

3. _____ TTL is the fastest within the TTL family.

Digital Circuits and Devices

Answers:

1. standard, high-speed, low-power, Schottky, low-power Schottky
2. clamping diodes, output impedance
3. Schottky

5-6 Nonsaturated Bipolar Logic Families

CURRENT-MODE LOGIC (CML)

Logic devices using transistors can be designed to work in the linear region of the transfer curve. Operated this way, collector current never reaches saturation. This eliminates the delay induced by a saturated transistor when recovering from saturation. As a result, the major advantage of using nonsaturated transistors as logic gates is the establishing of faster switching between the two logic levels.

One kind of logic device offering linear operation is called *current-mode logic* (CML). An example of a CML gate is shown in **Figure 5-16**. The transistor current regulator sets a nonsaturated operating point for Q_1 and Q_2. This regulator keeps the collector current of Q_1 and Q_2 from ever reaching the knee of the transfer curves of these logic transistors.

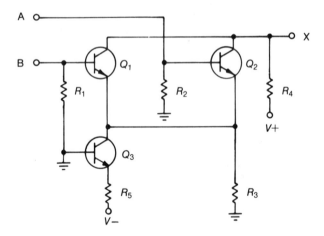

Figure 5-16. CML NOR gate

EMITTER-COUPLED LOGIC (ECL)

Another member of the nonsaturated bipolar logic group is the *emitter-coupled logic (ECL) family.* Figure 5-17 shows the emitter connections that character-

Logic Families

ize this family of logic gates. Notice that the schematic for the ECL gate shown resembles that of a circuit configuration for a differential amplifier.

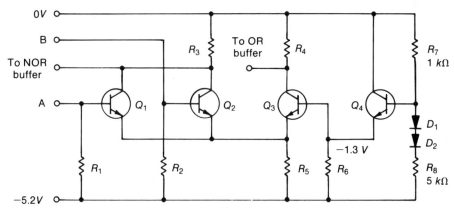

Figure 5-17. Example of an ECL gate

In this instance, one input of the "diff amp" is held at a fixed voltage by a regulator stage. R_7, D_1, D_2, and R_8 form a voltage divider at the base of Q_4, establishing an operating value of $-0.7\ V$. Since the emitter-base drop in Q_4 must be $0.6\ V$, this assures a constant $-1.3\ V$ *across* R_6 and at the base of Q_3 which is one side of the diff amp.

Q_1 and Q_2 are parallel inputs to the other side of the diff amp. Assuming positive logic, logic 0 for the ECL gate is $-1.6V$ (more negative) and logic 1 is $-0.8\ V$ (less negative). Logic 1 at A drives the base of Q_1 more positive, which turns on Q_1. Collector voltage on Q_1 moves farther from 0 V, making the collector more negative. This is logic 0. The same thing would happen if logic 1 input were applied to B. Since the output has been inverted, you'll recognize this as a NOR function.

At the same time, the voltage at the emitters has moved in a more positive direction—away from the $-5.2\ V$ supply value, because of the drop across R_5. With the base of Q_3 clamped by the regulator, a more positive voltage at the emitter results in a weaker collector current. The drop across R_4 is less, and the voltage at the collector of Q_3 moves in a less negative direction. This is the equivalent of logic 1, and logic hasn't been inverted. The function is OR.

Operation of these transistors cannot occur on the nonlinear portion of their transfer curves. Hence, switching can be almost instantaneous. ECL offers the fastest switching of any logic family, reaching 1000 *MHz* in some devices. However, power dissipation of ECL is five to ten times that of TTL.

INTEGRATED-INJECTION LOGIC (IIL or I²L)

At first glance, the circuitry of an I²L (integrated-injection logic) gate appears to

115

Digital Circuits and Devices

be nothing more than oversimplified TTL—no resistors, no diodes. However, transistors take the place of pull-up resistors at the inputs. This cuts power dissipation, allows high switching speeds, and renders IC fabrication less complicated than in virtually any other family of logic. For these reasons, I²L has become a favorite in certain large-scale integration (LSI) designs. **Figure 5-18** diagrams a simple NOR function of I²L construction.

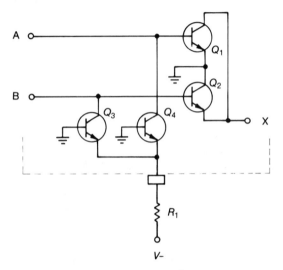

Figure 5-18. Example of I²L NOR gate

Note the open-collector arrangement of logic transistors Q_1 and Q_2, and also that R_1 isn't part of the chip. Current consumption of the gate can be decreased by raising the value of R_1 although speed of switching suffers. Operation is similar to ECL and CML.

You can't buy individual I²L gates, since their extremely high packing density makes them practical only for highly miniaturized, dedicated LSI packages.

Topic Review   5-6

1. The major advantage of using nonsaturated bipolar logic transistors is the establishing of _____ _____.

2. The three logic families of the nonsaturated bipolar logic group are _____, _____, and _____.

3. _____ offers the fastest switching of any logic family.

Answers:

1. faster switching
2. CML, ECL, I²L
3. ECL

5-7 MOS Logic Families

NMOS AND PMOS

MOS fabrication results in the channel being dominated either by negative carriers (*N-channel* MOS) or by positive carriers (*P-channel* MOS). Additionally, there are two modes of MOS operation. *Enhancement-mode* describes those devices in which zero bias leaves almost no current flowing in the channel—in other words, the cutoff condition. The symbol for an enhancement-mode device **Figure 5-19** depicts the channel element as a dashed line. An arrow pointing in toward the substrate element indicates *N*-channel; an arrow pointing out represents *P*-channel. Note that the corner of the gate symbol is placed near the preferred source connection, since the source and drain are interchangable.

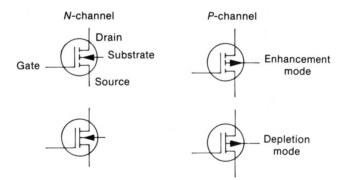

Figure 5-19. Enhancement and depletion mode MOS devices

For *N*-channel enhancement-mode MOS, a positive gate voltage causes conduction in the channel. With *P*-channel enhancement-mode, a negative bias is required at the gate to start the channel conducting.

Depletion-mode MOS devices—symbolized in schematic diagrams by a solid line as the channel element—conduct heaviest when no bias is applied to the gate. Bias voltage acts as a diminishing influence, determining how much current flows in the channel. Negative gate voltage "pinches off" channel current in an *N*-channel depletion-mode MOS transistor. Positive bias voltage controls or "pinches" channel current in *P*-channel depletion-mode MOS devices.

Digital Circuits and Devices

NMOS and PMOS transistors form the same kind of logic gates as bipolar transistors. MOS gates are slower though, and usually require higher voltage levels for logic 1. Otherwise, they behave virtually the same.

Of course, in schematics, MOS transistors have *drains* instead of collectors, *sources* in place of emitters, and *gates* instead of bases.

LOW THRESHOLD PMOS

In low threshold PMOS, the input and the output FETs have been replaced by silicon gates, which will increase the speed and allow easier interfacing to TTL circuits.

VMOS

In VMOS the gate region is cut in a "V" shape which lowers the gate resistance and allows the FET to switch at speeds three times that of standard NMOS or PMOS.

DMOS

In Double-diffused doping MOS, the substrate has special doping which allows the gates to dissipate about one-half the power as standard MOS, but the doping slows down the switching speed.

HMOS

In High performance MOS special doping and shapes of the FETs allow this family of MOS logic to be very fast and consume very little power, but the cost of this type of MOS is two to three times higher than any other type of MOS.

CMOS

Complementary MOS combines NMOS and PMOS transistors into a special kind of logic-circuit device. The hookup in **Figure 5-20a** uses four enhancement-mode MOS transistors—two N-channel and two P-channel. One advantage is the elimination of internal resistive components, thus improving speed and efficiency. Best of all, CMOS can be fabricated on a substrate in high density and at low cost.

As a final exercise to round out your study of logic families, try to figure out what kind of logic this CMOS gate performs. Take it one step at a time, filling in the truth table in **5-20b**.

Remember that Q_1 and Q_3 are *P*-channel enhancement-mode, and therefore carry current in their channels when logic applied to them is negative or logic 0. Q_2 and Q_4,

5
Logic Families

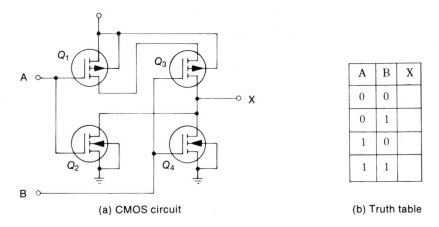

(a) CMOS circuit

A	B	X
0	0	
0	1	
1	0	
1	1	

(b) Truth table

Figure 5-20. Example of CMOS gate

on the other hand, are *N*-channel enhancement-mode, so their channels conduct when logic 1 is applied. Note also that the channels of Q_1 and Q_3 are connected in series. These clues should help you analyze this gate correctly.

Both inputs at logic 0: Q_1 and Q_3 both conduct. Q_3 and Q_4 remain cut off. V+ finds a ready path to output X through Q_1 and Q_3. Meanwhile, Q_2 and Q_4 are open, and can have no effect on output X. So, output X provides logic 1. Write that into the truth table.

Input A logic 0, input B logic 1: Q_1 remains on. However, the channel of Q_3 doesn't conduct, thus leaving the V+ path to X open. Q_2, being cut off, has no effect on the circuit. But Q_4 *does* conduct, pulling output X to ground or logic 0. Enter that into the truth table.

Input A logic 1, input B logic 0: This turns Q_3 on, and its channel is in the V+ path. But logic 1 at A leaves Q_1 off, so logic X still can't receive the positive voltage from V+. Q_4 has switched to cutoff, but now Q_2 conducts. Consequently, output X continues to be held at logic 0. Write logic 0 for X in the third line of the truth table.

Both inputs at logic 1: Now Q_1 and Q_3 both cut off, completely opening the V+ path to X. Logic 1 bias makes Q_2 and Q_4 both conduct, and they pull output X to logic 0. Make line 4 of the truth table show output X as logic 0.

Assessing the table you've completed, you should have no problem recognizing the truth pattern: this CMOS gate performs NOR logic.

CMOS SUBFAMILIES

The subfamilies of CMOS are standard CMOS, SOSCMOS, and HCMOS.

Digital Circuits and Devices

In SOSCMOS sapphire is used as an insulating material to the substrate, which will lower the reactance of the internal FETs, this will allow the speed to increase to the level of the slowest TTL sub-family. The major problem is the very high cost and the lack of all types of gates.

HCMOS is a high performance CMOS version.

Topic Review 5-7

1. The families of the MOS logic group are _____, _____, _____ _____ _____, _____, _____, _____, and _____.

2. CMOS combines _____ and _____ transistors to form a special kind of logic-circuit device.

3. The subfamilies of CMOS are _____, _____, and _____.

Answers:

1. NMOS, PMOS, Low threshold PMOS, VMOS, DMOS, HMOS, CMOS
2. NMOS, PMOS
3. regular CMOS, SOSCMOS, HCMOS

5-8 Summary of Logic Families
INTERFACING LOGIC FAMILIES

TTL to CMOS:
    Use a 10 $k$ohm pull-up resistor from the output of the TTL to the supply to bring the levels up to CMOS input levels.

CMOS to TTL:
    Normally requires special buffers or drivers, but most CMOS gates can drive one 74LS00 series gate input.

TTL to PMOS:
    When PMOS is operating off a $5V$ supply use a translator/buffer.

PMOS to TTL:
    When PMOS is operating off a $5V$ supply use a translator/buffer.

TTL to NMOS:
    When NMOS is operating off a $5V$ supply and ground you can use a $10k$ pull-up resistor.

NMOS to TTL:

5
Logic Families

When NMOS is operating off a +5V and −12V supply the NMOS will drive one 74LS00 series gate input.

TTL to ECL or CML:
  Must use a TTL to ECL translator/buffer.

ECL or CML to TTL:
  Must use a ECL to TTL translator/buffer.

TTL, NMOS, CMOS to IIL:
  Add current limiting resistor to input of IIL gate.

IIL to TTL, NMOS and CMOS:
  Will drive one input if pull-up resistor to supply is used.

USING LOGIC FAMILIES

Using TTL
1. Tie all unused inputs to ground or V+.
2. Do not tie standard TTL outputs together except when they are open-collector or tri-state, and make sure no more than one is enabled at any one time.
3. Maximum V+ should not exceed 7V.
4. Maximum input signal should not exceed 5.5V at a V+ of 5V.
5. Fan out in general is 10 within its own sub-family.
6. Install a $0.1\mu F$ to $0.01\mu F$ bypass capacitor between V+ and ground for every five chips.

Using MOS
1. Store in conductive foam.
2. Ground to earth when soldering.
3. In low-humidity conditions, ground your wrist to earth with 1 Mohm.
4. Do not remove or insert chip with power on.
5. Do not apply a signal to an input if the supply is off.
6. Connect all unused inputs to V+.

Table 5-6.

Typical characteristics of logic families

	RTL	DTL	TTL	HTL	ECL/CML	IIL	NMOS	PMOS	CMOS
Logic 0 (VDC)	0 to .4	0 to .8	0 to .8	0 to 5	−1.6	.1	0 to 2	−5 to −15	2 to 12
Logic 1 (VDC)	2 to 5	2 to 5	2 to 5	10 to 15	−.75	.6	5 to 12	0 to −2	1 to 4
Supply Level (VDC)	5	5	5	15	−5	1	12	−15	3 to 18
Power Dissipation (mW)	15	10	10	40	30	.2	2	2	.1
Propagation Delay (nsec)	50	25	10 to 30	70 to 85	1 to 5	25 to 250	30 to 60	30 to 70	20 to 50
Noise Margin (VDC)	.2	.8	1	6	.5	.4	.4 to 1/2 V+	(same)	(same)
Fan-out	4	8	10 to 15	10	20 to 30	10 to 45	20 to 50	(same)	(same)

Digital Circuits and Devices

5-9 Negative Logic

Where negative voltages operate the device, as in PMOS and ECL technology, it would seem natural for the more negative value to assume the logic 1 role. This leaves logic 0 nearer zero voltage, and therefore *less negative* than logic 1. Therefore, occasions arise where the concept of *negative logic* serves better than natural or *positive logic*. To avoid confusion, let's briefly review the fundamental meanings of digital logic.

Logic conditions are properly described as either true or false. Conventionally, logic 1 symbolizes the condition of logic true, and logic 0 indicates false. Moreover, logic true typically signifies an expected action, while logic false suggests some action that does not or did not occur.

Polarity enters the picture here, because in electronic circuits, the logic conditions 1 and 0 consist of DC voltages. Low voltage could be assigned the role of logic 1, with a higher value made logic 0. The result would be negative logic-that is, logic 1 would be more negative (which is to say less positive) than logic 0.

Most concepts of digital circuitry today adhere to the positive logic convention. Nevertheless, you have to be familiar with negative logic, since designers sometimes choose to portray negative-going logic as the activating influence in a circuit. If it is, negative logic gates will appear in the circuit diagram. You can recognize them by the not-circles at all terminals, inputs as well as outputs.

NEGATIVE-LOGIC GATES

Study the two gate diagrams in **Figure 5-21**. They resemble typical AND and OR gates. However, not-circles at inputs and outputs identify them as negative AND and negative OR gates.

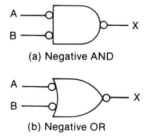

(a) Negative AND

(b) Negative OR

Figure 5-21. Symbolic representation of negative logic gates

You can view such gates in either one of two ways: as negative-logic gates or as positive-logic gates. Which one you choose will usually depend on the design of the equipment and on the nature of its power supplies. A negative-logic approach may seem better when the DC supply hookup places ground positive and "hot" negative—

5
Logic Families

as with PNP or PMOS devices.

Let's study the negative AND gate in **Figure 5-21a.** Remember for negative-logic concepts, logic 1 is always negative-going; negative logic to both inputs turns the gate on, just as positive logic would activate an ordinary (positive) AND gate. Output is logic 1, but in negative logic, if only one input of this gate saw negative-logic 1, the gate would remain off and the output would be logic 0 —negative-logic 0.

Negative OR works like ordinary OR, except for the reversed voltage relationships. Negative logic 1 at either input in **Figure 5-21b** turns the gate on, making output logic 1. Remember that with a negative OR gate, the logic 1 output is always a more negative voltage than logic 0.

This sums up the whole idea. As long as you remember that logic 1 and logic 0 are only symbols, and *not* actual voltage levels, you'll be able to work with negative-logic systems as easily as with any other.

POSITIVE LOGIC THROUGH NEGATIVE GATES

Perhaps you've already recognized that concepts of positive logic work with negative-logic gates. You know that, in positive electronic logic, not-circles signify inversion wherever they appear. You might find it easier, therefore, to always trace circuit action in terms of positive logic, whatever the gate designs.

Consider the negative AND gate in **Figure 5-21a** in terms of positive logic. A not-circle inverts the logic 1 reaching input A. If you think of the AND gate in ordinary terms, you know that the resulting logic 0 leaves the gate inhibited, no matter what happens at the other input. The output of an inhibited AND is logic 0. Here, the output not-circle inverts the logic, making it logic 1.

For negative OR, logic 0 at either input turns the gate on, since an input not-circle inverts the logic before the gate itself "sees" it. The output of a regular OR gate would be logic 1 when the gate enables; however, this gate also has a not-circle, which inverts the output. So, logic 0 fed to any input causes logic 0 output.

NEGATIVE-LOGIC TRUTH TABLES

Let's examine some truth tables that will help with these explanations. They'll confirm that negative AND gates and negative OR gates figure out the same, regardless of whether negative or positive logic is used. **Figure 5-22** contains truth tables done both ways. They prove that the terms logic 1 and logic 0 mean the same thing in both negative and positive logic. *Only the voltage levels that they symbolize are different.* Test each truth table now, by working out all the input combinations for yourself.

123

Digital Circuits and Devices

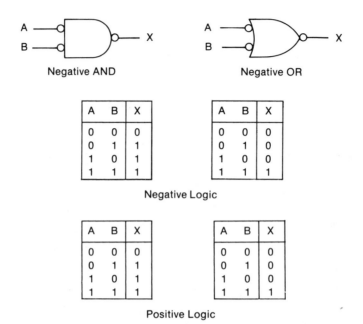

Figure 5-22. Examples of negative logic gates with their truth tables

Negative gates are considered *low-active* devices. In digital electronics, a low-active device is one in which logic 0 causes the expected activity: that is, logic 0 is enabling, while logic 1 inhibits. This low-active concept was used to form the negative OR truth table in **Figure 5-22.**

Logic 0 at either input enables the negative gate, so that the output is logic 0. Logic 0 at only one input—either one—enables the gate, and the output is still logic 0. Logic 1 at both inputs inhibits the gate, for an output of logic 1.

Try working out the table for negative OR by positive logic. Logic 0 at both inputs inverts to logic 1, enabling the gate; logic 1 output inverts to logic 0. Logic 1 at one input is inverted to logic 0, which has no effect on the gate; but logic 0 at the other input is inverted to logic 1, enabling the gate. Again, output inverts to logic 0. Logic 1 at both gates inverts to logic 0, and the gate remains inhibited: its logic 0 output inverts to logic 1.

Conclusion: You can assess the operation of a negative logic gate in terms of either negative or positive logic—whichever is easier for you. It's only when you become concerned with logic *voltage* levels that the difference matters.

One recommendation: Form the habit of thinking in terms of positive logic—for two reasons. Most systems incorporate mainly positive logic, and even those designed and diagrammed with negative-logic gates can be figured out by applying the principles of positive logic.

5
Logic Families

EQUIVALENT LOGIC IN DIAGRAMS

One more source of possible confusion needs clearing up. According to **Figure 5-22**, any negative-logic gate produces the same truth table, whether you reason by positive or negative logic. This doesn't mean, however, that negative gates function the same as their positive namesakes. A negative AND gate, for example, isn't the same as a positive AND device.

Again, truth tables bear this out. **Figure 5-23** compares AND gates of both polarities. Keep in mind the two prime differences: 1) the enable logic for negative-logic gates is logic 0, whereas it's logic 1 for positive versions; 2) negative-logic gates output logic 0 when enabled, whereas their positive versions output logic 1. You can see from the truth tables that the two gates definitely do *not* perform the same function.

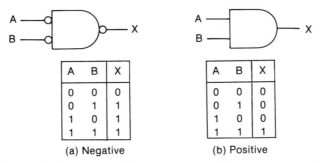

Figure 5-23. Comparision of negative and positive AND gates

Figure 5-24 compares negative OR and positive OR. Here, too, the primary differences in logic build truth tables that are the opposite of each other.

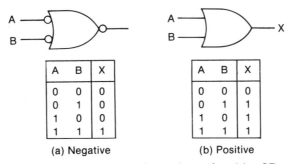

Figure 5-24. Comparison of negative and positive OR gates

Compare the negative AND truth table in **Figure 5-23a** with that for positive OR in **Figure 5-24b**. They're alike. The fact is, negative AND is the logical equivalent of positive OR. Truth tables will also show that negative OR and positive AND are logical equivalents.

125

Digital Circuits and Devices

Topic Review 5-9

1. Negative logic is sometimes used in the anlysis of logic gate operation when _____ _____ operate the logic devices.

2. The terms _____ and _____ mean the same thing in both negative and positive logic; however, the _____ _____ they symbolize are different.

3. Positive AND is the logical equivalent of _____ _____.

4. _____ _____ is the logical equivalent of positive OR.

Answers:

1. negative voltages
2. logic 1, logic 0, voltage levels
3. negative OR
4. negative AND

5-10 Summary Points

1. Logic gate functions can be made from a combination of bipolar transistors or diodes or metal-oxide-semiconductor FETs, and depending on which type and configuration is used will determine the type of logic family. Each logic family has certain advantages and disadvantages over other logic families, which family is picked will be determined by the characteristics needed.

2. The logic families within the saturated bipolar logic group are RTL (Resistor-Transistor Logic), DTL (Diode-Transistor Logic), HTL (High-Threshold Logic), and TTL (Transistor-Transistor Logic).

3. The subfamilies of TTL are standard, high-speed, low-power Schottky, and low-power Schottky TTL.

4. The logic families within the nonsaturated bipolar logic group are CML (Current-Mode Logic), ECL (Emitter-Coupled Logic), and IIL or I^2L (Integrated-Injection Logic).

5. The logic families with the MOS logic group are NMOS (N-channel MOS), PMOS (P-channel MOS), HMOS (high performance MOS), and CMOS (Complementary MOS).

5 Logic Families

6. The most important specifications for any logic families are, the type of family, compatibility with other families, operating voltages, power dissipation, logic levels, propagation delay, noise margin, temperature range and contents of the IC.

7. Logic Families in short:
   Largest power dissipation — HTL, ECL, CML.
   Smallest power dissipation — MOS logic group.
   Fastest — ECL, CML.
   Slowest — MOS, HTL, IIL.
   Largest noise margin — HTL.
   Smallest noise margin — RTL.
   Largest fan-out — MOS, IIL.
   Smallest fan-out — RTL.
   Most used logic family SSI — MSI — TTL.
   Most used logic family LSI — VLSI — MOS.
   Most used logic gate — NAND, NOR.

8. Logic gates work on voltage levels. In negative logic the voltage level for a logic 0 is zero volts, and for a logic 0 the voltage level is the supply voltage. A negative logic AND acts as a positive OR gate, also a negative OR acts as a positive logic AND gate.

5-11 Chapter Progress Evaluation

1. The word _____ describes the condition reached by a transistor when its collector current has risen to the maximum value. Such operation gives rise to what is called _____ logic.

2. Which element of a transistor is used as the input for most TTL logic devices?

3. The fastest TTL device is _____.

4. True or False? ECL and CML devices belong to a family group called nonsaturated logic, because they operate in a linear portion of the transfer curve.

127

5. Next to each word at the right, print the letter that indicates the matching definition.

   A. High positive voltage turns the channel on.   _____Complementary MOS

   B. It combines PMOS and NMOS on a single substrate.   _____Depletion MOS

   C. Channel carries high current with no bias.   _____Enhancement NMOS

   D. Positive bias leaves no current in the channel.   _____Depletion PMOS

6. Schottky diode clamping is a technique for speeding up logic gates. Which family of logic devices uses this clamping?

7. Which family of logic devices operates the fastest?

8. What range of operating temperatures is considered a typical specification for logic devices in an industrial environment?

9. Which component is built into the input of an HTL device?

10. Which of the following is true of logic devices of the MOS family?

    A. They offer faster switching times than all families except ECL.

    B. They draw more power-supply current than any other type.

    C. They can't be combined on a single substrate.

    D. They operate in both the enhancement and depletion modes.

Chapter 6

Digital Simplification

Objectives

Upon completion of this chapter, you should be able to do the following:

* Define Boolean algebra

* Explain the relationship between Boolean identities and the operation of digital gates

* List and explain the five basic Boolean laws governing variables

* Decribe the purpose and use of DeMorgan's law

* Describe the purpose of Karnaugh mapping

* Simplify any size Boolean expression, using Karnaugh mapping and Boolean algebra

6-1 Introduction

The English mathematician George Boole developed a system of theorems and postulates for *yes/no* logic around 1850. His ideas brought together the principles of logic and mathematics, producing the discipline now called *Boolean algebra*.

In 1938, a researcher at M.I.T. (Massachusetts Institute of Technology) demonstrated the value of Boolean algebra in designing digital electronic circuitry. Since then, experts have both expanded and simplified Boole's logic methods as they apply to electronics, and digital circuitry has come to be called logic circuitry.

From your studies of Boolean concepts earlier you may recall that AND, OR, and NOT are the fundamental logic operators. Everything in digital circuitry uses some combination of these basic functions. As you approach the world of electronics there are a few more facts you need to know about Boolean operators.

Digital Circuits and Devices

Boolean algebra simplification is divided into three sections:
1. Boolean identities.

2. Boolean laws of algebra.

3. Boolean theorems.

6-2 Boolean AND Identities

Boolean AND identities will show the relationship between the inputs of an AND gate and the output condition of the AND gate. AND identities allow the user to reduce the AND gate circuit to a piece of wire with some type of level or variable connected to it. The most important thing to remember is how an AND gate works. The AND gate will produce a *high* on its output only when all its inputs are *high*. At all other times the output will be *low*.

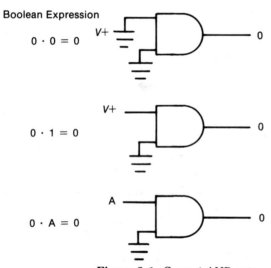

Figure 6-1. Group 1 AND gate

As can be seen from the examples in Group 1, each expression has at least one input to the AND gate grounded, which will produce a *low* on its output no matter what level is on the other input. So if any input to an AND gate is *low* the output will *always be low* and can be replaced by a wire that is grounded.

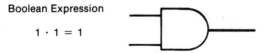

Figure 6-2. Group 2 AND gate

6
Digital Simplification

As can be seen from the example in Group 2, the AND gate expression has all of its inputs tied to V+, which will always produce a *high* on its output. So the gate can be replaced by a wire that is tied *high*.

Boolean Expression

1 · A = A

A · A = A

Figure 6-3. Group 3 AND gate

As can be seen from the examples in Group 3, each expression is controlled by the level of variable "A". If "A" is *high* the output will be *high*, if "A" is *low* then the output will be *low*. So the output condition of the circuit will be controlled by the value of "A" and "A" alone. A wire with "A" connected to it will work the same way.

Boolean AND identities:

0 · 0 = 0

0 · 1 = 0

0 · A = 0

1 · 1 = 1

1 · A = A

A · A = A

Before going on be sure you understand how all the AND identities relate to the operation of an AND gate.

Topic Review   6-2

1. The AND gate will produce a _____ on its output only when all of its inputs are *high*.
2. If any input to an AND gate is *low*, the output will always be _____ and can be replaced by a wire that is grounded.

Digital Circuits and Devices

Answers:

1. *high*
2. *low*

6-3 Boolean OR Identities

Boolean OR identities will show the relationship between the inputs of the OR gate and the output condition of the OR gate. OR identities allow the user to reduce the OR gate circuit to a piece of wire with some type of level or variable connected to it. The most important thing to remember is how an OR gate works. The OR gate will produce a *high* on its output whenever any of its inputs are *high*. The only time the output goes *low* is when all its inputs are *low*.

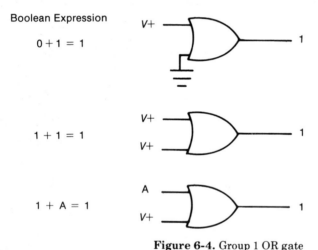

Figure 6-4. Group 1 OR gate

As can be seen from the examples in Group 1, each expression has at least one input to the OR gate tied to supply, which will produce a *high* on its output no matter what level is on the other inputs. In short if any input to an OR gate is *high* the output will *always be high* and can be replaced by a wire that is connected to the supply.

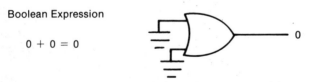

Figure 6-5. Group 2 OR gate

As can be seen from the example in Group 2, the OR gate expression has all its inputs grounded, which will always produce a *low* on its output. So the gate can be replaced by a wire that is grounded.

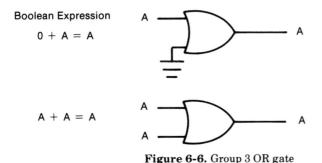

Figure 6-6. Group 3 OR gate

As can be seen from the examples in Group 3, each expression is controlled by the level of variable "A". If "A" is *high* the output will be *high*, if "A" is *low* then the output will be *low*. So the output condition of the circuit will be controlled by the value of "A" and "A" alone. A wire with "A" connected to it will work the same way.

Boolean OR identities:

0 + 1 = 1

1 + 1 = 1

1 + A = 1

0 + 0 = 0

0 + A = A

A + A = A

Before going on be sure you understand how all the OR Boolean identities relate to the operation of an OR gate.

Topic Review 6-3

1. The OR gate will produce a _____ on its output whenever any one of its inputs are *high*.

2. The only time the output goes _____ is when all of its inputs are *low*.

Digital Circuits and Devices

Answers:

1. *high*
2. *low*

6-4 Boolean NOT Identities

Boolean NOT identities will show the relationship between the input and output of an inverter. NOT identities allow the user to reduce the inverted circuit to a piece of wire with some type of level or variable connected to it. The most important thing to remember is how the inverter works. The inverter will produce the opposite condition on its output than the condition on its input.

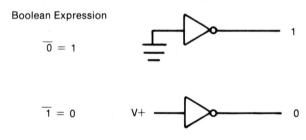

Figure 6-7. Group 1 NOT gate

As can be seen from the examples in Group 1, the output of a NOT function will always have the opposite condition than the level on its input. So you could use a piece of wire connected to the output level and the circuit will work the same.

Boolean Expression

$\overline{\overline{A}} = A$

$\overline{A} \longrightarrow\!\!\rhd\!\circ\!\longrightarrow X = \overline{[\overline{A}]} = A$

Figure 6-8. Group 2 NOT gate

As can be seen from Group 2, when $\overline{A}$ is applied to the input of an inverter the output will be the opposite level which is A.

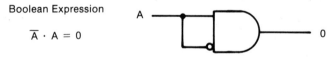

Figure 6-9. Group 3 AND gate NOT input

As can be seen by Group 3, if an AND gate has an input that is also inverted, one input will always be *low* and one always *high*. This condition will cause the output to produce a *low* all the time. So a wire that is connected to ground will work the same.

6
Digital Simplification

Boolean Expression

$\overline{A} + A = 1$

Figure 6-10. Group 4 OR gate NOT input

As can be seen by Group 4, if an OR gate has an input that is also inverted, one input will always be *low* and one always *high*, which will produce a *high* on its input. So a wire that is connected to supply will work the same.

Boolean NOT identities:

$\overline{0} = 1$

$\overline{1} = 0$

$\overline{\overline{A}} = A$

$\overline{A} \cdot A = 0$

$\overline{A} + A = 1$

Before going on be sure you understand how all the NOT Boolean identities relate to the operation of an inverter.

Topic Review 6-4

1. The inverter or NOT gate will produce the _____ on its output than the condition on its input.

2. If an AND gate has an input that is inverted, one input will always be _____ and one always _____, producing a *low* at all times.

3. If an OR gate has an input that is inverted, one input will always be _____ and one always _____, producing a *high* on its output.

Answers:

1. *opposite*
2. *low, high*
3. *low, high*

Digital Circuits and Devices

6-5 Boolean Laws

In Boolean algebra, gate reduction is performed by the use of algebraic laws and identities. The Boolean laws allow the user to rearrange a Boolean expression without changing its function. There are five laws and they are as follows:

1. Associative law.

2. Commutative law.

3. Associative and Commutative law.

4. Distributive law.

5. DeMorgan's law.

ASSOCIATIVE LAW

The associative law allows the user to change the groupings of variables, without changing the operation of the circuit. Grouping can only be changed if all inputs belong to the same logic function. For example:

$$[B + C + D] = [B + [C + D]] = [[B + C] + D]$$

$$[A [BC]] = [ABC] = [[AB]C]$$

In both examples the circuit will operate the same. The only difference is the number of gates required for each circuit. If you have any doubt make a truth table for each expression.

COMMUTATIVE LAW

The commutative law allows the user to rearrange the order of any variable in a gate circuit. The rearrangement of variables can only occur within the same logic function. As an example:

$$[A + E + B] = [A + B + E]$$
$$= [B + E + A]$$
$$= [E + B + A]$$
$$= [B + A + E]$$

136

6
Digital Simplification

$$= [E + A + B]$$

$$[ABC] \quad = [BCA]$$

$$= [BAC]$$

$$= [CBA]$$

$$= [CAB]$$

$$= [ACB]$$

In both examples the circuits will operate the same, the only difference is how the gate inputs are arranged. If you have any doubt make a truth table for each expression. Note this law alone will not reduce the number of gates a circuit has but allows the user to rearrange the order of variables which can be used with other laws.

ASSOCIATIVE AND COMMUTATIVE LAWS

When the associative and commutative laws are used together, the grouping and order of the variables can be changed without affecting the circuit's operation. As an example:

$$[A + C + B] = [[A + B] + C]$$

$$= [[B + C] + A]$$

$$= [[C + B] + A]$$

$$= [[C + A] + B]$$

$$= [[B + A] + C]$$

$$= [A + B + C]$$

$$[[BE]F] \quad = [BEF]$$

$$= [EFB]$$

$$= [EBF]$$

$$= [[EB]F]$$

137

Digital Circuits and Devices

$$= [[FB]E]$$

In both examples the circuits will operate the same, the only difference is the order of the variables. If you have any doubts make a truth table for each expression.

DISTRIBUTIVE LAW

The distributive law allows the user to rearrange the gates and still obtain the same logic function. As an example:

First distributive law states:

$$A \cdot [B + C] = [A \cdot B] + [A \cdot C]$$

And second distributive law states:

$$A + [B \cdot C] = [A + B] \cdot [A + C]$$

The circuits in both examples will operate the same but the way the circuits are connected together are very different. If you have any doubt derive a truth table for each Boolean expression.

DEMORGAN'S LAW

Identifies for Boolean algebra only affect the AND, OR and NOT logic functions, but in digital electronics there are also NOR and NAND logic functions. DeMorgan's law is used to convert a NOR and NAND logic function into an AND and OR logic function with inverted inputs. There are four steps in DeMorgan's law when performed on a NOR and NAND gate. They are as follows:

1. Cancel any double negation lines over the entire Boolean expression. If this cancels all the negation lines above the expression, the law is complete. If a negation line still remains, continue on with the following steps.
2. Change the sign of each logic function.
3. Invert each input.
4. Cancel any double negation lines over any input.

As four practical examples, DeMorgan's law states

$$\overline{A \cdot B} = \overline{A} + \overline{B}$$

$$\overline{\overline{AB}} = AB$$

$$\overline{A + B} = \overline{A} \cdot \overline{B}$$

$$\overline{\overline{\overline{AB}}} = \overline{A} + \overline{B}$$

6
Digital Simplification

Topic Review   6-5

1. The associative law allows the user to change _____ of _____ without changing the operation of the circuit.

2. The commutative law allows the user to rearrange the _____ of any variable in a gate circuit.

3. The distributive law allows the user to rearrange the _____ and still obtain the same logic function.

4. DeMorgan's law is used to convert _____ and _____ logic functions into _____ and _____ logic functions with inverted inputs.

Answers:

1. groupings, variables
2. order
3. gates
4. NAND, NOR, OR, AND

6-6 Boolean Theorems

Boolean theorems are conditions that occur in Boolean expressions often. Since these conditions occur so often, if you remember how these theorems work and can recognize them in a Boolean expression, Boolean algebra will be easier to work with.

THEOREM 1 (COMMON VARIABLE REDUCTION)

$$A + [AB] = A$$

As a proof:

$$A + [AB] = A[1 + [1B]] = A[1 + B] = A[1] = A$$

Note the number 1 is placed in the locations where the variable has been removed to represent that the variable was there. Then use Boolean identities to reduce the problem further.

Here is some additional proof that the reduction of common variables works.

Digital Circuits and Devices

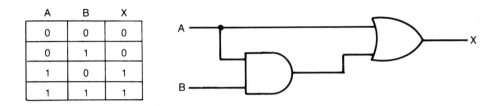

Figure 6-11. Truth table equivalent

As can be seen from the truth table in **Figure 6-11**, input A controls the output of the circuit, so a wire with A connected to it will function the same way. Boolean algebra uses the fact that one circuit with less gates or inputs will function the same way as a larger circuit with more inputs. This theorem will work with single and multi-variable expressions as long as the variables are in both gates and are ORed together. The identical variable will control the output and any other variable will have no effect on the output.

Other examples of THEOREM 1 :

[AB] + [ABC] = AB

A + [ABEF] = A

THEOREM 2 (COMPLEMENT VARIABLE REDUCTION)

$$A + [\overline{A} \cdot B] = A + B$$

As a proof:

$$[A + \overline{A}][A + B] = [1][A + B] = A + B$$

Note the complement of the single variable drops out of the expression. This theorem works only when a complement of a single variable is ORed to an AND gate.

Here is some additional proof of complement reduction of an expression:

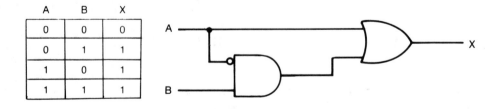

Figure 6-12. Truth table equivalent

6
Digital Simplification

As can be seen from the truth table in **Figure 6-12**, the truth table represents the OR function with A and B as inputs. In this theorem the complement of the single variable drops out of the expression. This theorem will only work with single variables ORed to the complement of the single variable.

Other examples of THEOREM 2:

$$\overline{A} + [AB] = \overline{A} + B$$

$$A + [\overline{A} \cdot \overline{B}] = A + \overline{B}$$

Now that Boolean identities, laws, and theorems have been studied, let's start reducing Boolean expressions. There are no rules for reducing Boolean expressions but here are some hints:

1. "DeMorganize" any inverted logic function.

2. Look for single variables and use Boolean laws to rearrange the expression so that identities and theorems can be used to reduce the number of gates and inputs a circuit has.

3. Reduction is complete when the number of gates cannot be reduced and the fewest numbers of inputs are in the circuit.

Remember, Boolean algebra reduces the number of gates and inputs a circuit has and will yield a smaller circuit that will operate the same way. If the Boolean expression is in lowest terms reduction will not be possible. If this is the case, the original expression is the one to use.

The following examples show Boolean algebra reductions:

Example 1 :

A[A + [$\overline{A} \cdot \overline{B}$]]	(5 gates - 2 AND, 1 OR, 2 NOT)
A [A + $\overline{B}$]	(theorem 2)
[A · A] + [A · $\overline{B}$]	(distributive law)
A + [A · $\overline{B}$]	(AND identity)
A	(theorem 1)

therefore: A[A + [$\overline{A} \cdot \overline{B}$]] = A   (requires 0 gates)

141

Digital Circuits and Devices

Example 2 :

$[\overline{A} + B + C] \cdot \overline{B} \cdot \overline{C}$ (5 gates - 1 AND, 1 OR, 3 NOT)

$[\overline{A}\,\overline{B}\,\overline{C}] + [B\overline{B}\,\overline{C}] + [C\overline{B}\,\overline{C}]$ (distributive law)

$[\overline{A}\,\overline{B}\,\overline{C}] + 0 + 0$ (NOT identity)

$\overline{A + B + C}$ (DeMorgan's Law)

therefore: $[\overline{A} + B + C] \cdot \overline{B} \cdot \overline{C} = \overline{A + B + C}$ (requires 1 gate)

One final example includes DeMorgan's law again:

$[\overline{ABC}] + [\overline{A + B}]$ (5 gates - 1 NAND, 1 NOR, 1 OR, 2 NOT)

$[\overline{A} + \overline{B} + \overline{C}] + [\overline{\overline{A}\,\overline{\overline{B}}}]$ (DeMorgan's law)

$[\overline{A} + \overline{B} + \overline{C}] + [AB]$ (Not identity)

$\overline{A} + [AB] + \overline{B} + \overline{C}$ (Associative and Commutative law)

$\overline{A} + B + \overline{B} + \overline{C}$ (Theorem 2)

$\overline{A} + 1 + \overline{C}$ (OR identify)

$\overset{1}{\overline{}}$

therefore: $[ABC] + [\overline{A + B}] = 1$ (requires 0 gates, output will always be *high*)

Topic Review   6-6

1. Theorem 1, common variable reduction, states _____.

2. Theorem 2, complement variable reduction, works only when a _____ of a single variable is ORed to an _____ gate.

Answers:

1. $A + [AB] = A$
2. complement, AND

6
Digital Simplification

6-7 Karnaugh Mapping

Simplification has high priority in logic circuit design. The chief goal, after all, is to *minimize* the number of logic gates needed to accomplish a specific function. (This reduces circuitry and costs.) Boolean algebra is one effective tool for accomplishing this. But lengthy equations can be complicated to minimize; Boolean algebra may even miss certain minimization possibilities.

A Karnaugh map can do the same thing: that is, find the most compact logical equivalent of a complex truth table or Boolean expression. **Karnaugh mapping supplies a highly visual way of simplifying and minimizing logic functions.**

Essentially, a Karnaugh map is a grid whose coordinates represent the various terms of a Boolean-type expression. A simple map form, for mapping expressions with two digital-logic variables, appears in **Figure 6-13(b)**. This map charts what are called *minterms* —ANDed variables.

Figure 6-13 also shows a truth table (6-13c) and Boolean equation (6-13a) for the function of the production term A B, or simply AB. These represent the simplest expression possible with two variables; one minterm, X = AB is so simple that it needs no map, actually. But we will use it to study the correlation between a Boolean expression and a Karnaugh map (and between a truth table and the map.)

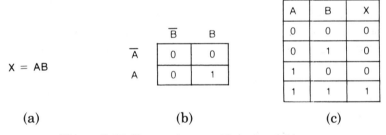

Figure 6-13. Karnaugh map with two variables

Look first at the Boolean equation. It states that X proves true only if both A and B are true. The table states the same truth, on its bottom line.

The difference is, the truth table includes all the other possible conditions of A and B. The Karnaugh map, (sometimes called a K-map), does this, too, by means of its coordinates. For conditions when A is logic 0, an A-not row has been provided. A B-not column allows for B input at logic 0.

A solution for each combination appears in the appropriate square of the K-map. Accordingly, for A-not and B-not (which means, of course, that A and B are both *low*), the solution (output X) is logic 0. For A-not and B, output is logic 0; for A and B-not, again 0. Finally, you'd read the lower right square as "X is 1 when A and B are 1."

The larger Karnaugh map in **Figure 6-14** illustrates the mapping scheme for a

Digital Circuits and Devices

Boolean equation with three variables. Keep in mind that each variable can exhibit either of two states. Thus, a grid capable of encompassing every possibility must have eight squares—the third power of 2. Accordingly, by this measure, a four-variable K-map calls for 16 squares.

The zeroes plotted in the map of **Figure 6-14** are not necessary. Logic 1 appears where A and B meet C. That one mark plots the solution for Boolean minterm ABC. No other conditions have to be mapped, only those that produce a logic 1 solution.

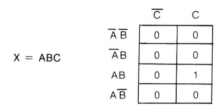

Figure 6-14. Karnaugh map with three variables

WHAT A KARNAUGH MAP CAN SHOW

A Karnaugh map really proves its value when mixed expressions must be solved. Consider the Boolean equation X = ABC + A$\overline{B}$C, a sum-of-products equation. A Karnaugh map can readily simplify this relatively complex equation best without recourse to the list of laws required in Boolean algebra. Each group of ANDed variables comprises a *minterm* that can be mapped. (It helps to assign a different number to each minterm, for identification later in the map.)

A Karnaugh map of this equation has been plotted in **Figure 6-15**. The Boolean expression states that X is true (logic 1) if A and B and C are true, or if A and C are true and B is false.

For more than one minterm, use each identifying numeral as if it means logic 1. Thus, a numeral goes into the map grid wherever the coordinates match its minterm. You place a 1 (corresponding to the first minterm) in the one square at which A, B, and C are all *high*. A 2 (for the second minterm) goes in the square below—the only place where *high* A, *low* B, and *high* C come together.

Since the two numerals occupy adjacent grids, they can be *looped*, as shown by the oval that's been drawn to enclose them in **Figure 6-15.** Inside the loop, one variable *complements*—that is, changes polarity—and therefore cancels. With B canceling B-not, only A and C are left. This is called the *loop value*. It is itself a minterm—a new one resulting from Karnaugh simplification.

In this example, there's only one loop, so its loop value becomes the only minterm in the new equation. Hence, the original expression reduces to X = AC.

144

6
Digital Simplification

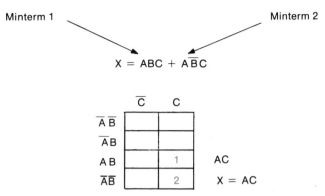

Figure 6-15. Karnaugh map comprising a minterm of ANDed variables

Remember that the variables A, B, and C represent logic inputs. The resulting equation states: "X proves true if A and C." In other words, X goes logic 1 if A and C see logic 1. A mildly complex logic function with ANDs ORed has been *minimized* to a simple AND.

TRUTH TABLE EQUIVALENTS

Take a look at the two truth tables in **Figure 6-16.** They state truths for the original expression (6-16a) and the new one (6-16b).

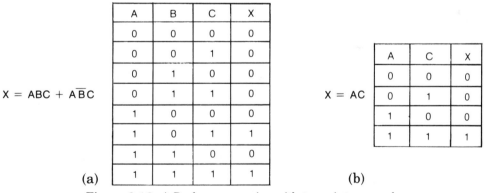

Figure 6-16. A Boolean expression with two minterms and three variables

According to the initial Boolean equation, two different sets of input conditions can bring output X to logic 1. A correct truth table verifies this, and shows that no other conditions permit logic 1 output. Logic 1 output occurs either when A and B and C are all *high*, or when A and C are *high* and B is *low*. In binary terms, X goes logic 1

145

Digital Circuits and Devices

only when the binary input pattern is either 101 or 111. All other conditions leave X at logic 0. So, the truth table at **6-16a** should have logic 1 in the output column opposite only the 101 and 111 rows. Every other input condition produces logic 0 for X.

Analyzing the lines of this three-input table, you can see that it makes no difference what logic input B sees. When A and C see logic 1, output X delivers logic 1. If either A or C fails to see logic 1, output X delivers logic 0—no matter what appears at B. Since B has no influence, input B may as well be omitted—and the truth table of **Figure 6-16b** does just that.

However, the Karnaugh map in **Figure 6-15** still offers three advantages over the truth tables in **Figure 6-16**: 1) the K-map uncovers that complementary B and B-not relationship more readily than a truth table can; 2) it eliminates any unnecessary inputs from the expression; and 3) a K-map is quicker to draw and easier to use than a truth table.

MINIMIZING LOGIC CIRCUITRY

Still another advantage of the Karnaugh map shows up when you prepare equivalent logic circuitry. Taken as is, the longer Boolean expression calls for the several gates in **Figure 6-17a**. Even after changing one AND gate to a type with a NOT input, you'd still need at least three gates to accomplish the function.

If you trace the logic in **Figure 6-17a**, you'll find that at least one AND gate turns on whenever both A and C see logic 1.

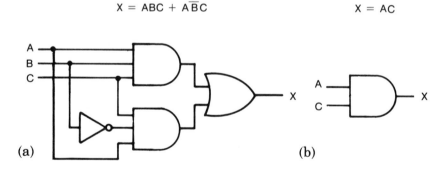

Figure 6-17. Simplifying a circuit using Karnaugh mapping

That's because B applies logic 1 to either one AND gate or the other, no matter which logic condition appears on input line B. When any AND gate enables, the OR gate does, too.

As for other input patterns, logic 0 at input A or input C inhibits both AND gates, thus inhibiting the OR gate. The output under those conditions is logic 0.

6
Digital Simplification

Figure 6-17b is wired according to the Karnaugh simplification: X = AC. You gain the same function, but with a single gate. Logic 1 appears at X when A and C both go high. (Otherwise X shows logic 0.) Since a B input has no effect anyway, there's no point in connecting one into the circuit.

All three methods of visualizing this logic function permit logic circuit minimization. However, the Karnaugh map usually wins out over the other two methods: it can be drawn quickly, and makes complementary relationships easy to see and cancel. As you deal with more complicated logic functions, involving a number of variables and consequently longer terms, you'll find Karnaugh mapping the one consistent way to simplify logic expressions.

DRAWING LARGER KARNAUGH MAPS

A Karnaugh grid contains enough squares to satisfy all logic combinations for the expression being mapped. Since each variable offers two possible states, the number of squares must equal 2 taken to whatever power corresponds with the number of variables to be mapped. Thus, two variables require four squares; three variables need eight; four variables, sixteen; and so forth.

Variables are arranged along the edges of the square or rectangle. Conventionally, the first variables go at the side, with the others along the top. **Figure 6-18** contains K-map forms for 2, 3, and 4 variables. Labels are included for the two possible states of every variable.

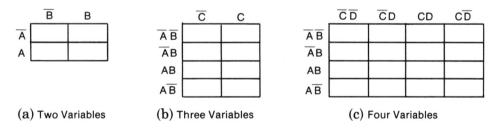

Figure 6-18. K-map forms for 2, 3, and 4 variables

Circuits requiring more than four inputs can sometimes be minimized by the use of a special trick—rearranging their Boolean expressions in such a way that you map no more than four variables at a time. (Only three-dimensional Karnaugh mapping can handle more than four variables in a single map.)

147

Digital Circuits and Devices

Use the forms in **Figure 6-18** as guides for up to four variables. Don't rearrange the labels; certain variants would work, but most would result in incorrect looping.

PLOTTING MULTIPLE VARIABLES

Imagine you need a digital decoder that turns on when fed any one of the eight decimal numbers listed in the left-hand column of **Figure 6-19a**. The problem: to find the minimum of combinational logic required for such a peculiar decoder.

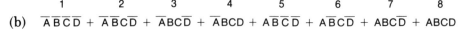

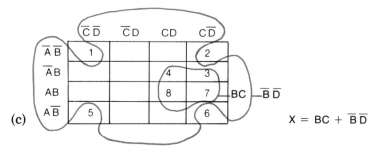

Figure 6-19. Plotting multiple variables

Start by converting the decimal values to binary patterns. The decimal-binary chart shows that it takes four binary digits to represent the highest number. Deduction: four inputs will be necessary. With such a variety of input patterns, minimizing by any method other than Karnaugh mapping would be extremely cumbersome. And you certainly don't want the circuitry to be cluttered with dozens of gates and inverters if you can help it.

6
Digital Simplification

Actually, the Boolean equation you begin with is relatively easy to write. Name a variable for each binary input position. To assign logic 0 as the required condition of a variable, place a not-bar over the variable name. Where the input condition must be logic 1, leave the letter plain.

Figure 6-19b demonstrates how the eight binary patterns were converted into eight logical products (minterms). The one output, X, must equate with any one of these minterms. To form a single Boolean expression, then, all the ANDed minterms must be ORed together. If you've worked out each minterm correctly, you should produce the Boolean equation shown at **6-19b**.

It should be clear that minimizing this lengthy list of minterms by means of Boolean theorems would be time-consuming and tedious. Instead, plot the minterms on a Karnaugh map.

Four variables call for a 16-square Karnaugh grid. To help you keep track of the minterms, give each one a number, as has been done in **Figure 6-19b.** Cover up the map plotted in **Figure 6-19c** and try creating your own on a separate sheet.

Minterm 1 fits into the top left corner, where coordinates come together for $\overline{A}\,\overline{B}\,\overline{C}$ and $\overline{D}$. Minterm 2 fits where $\overline{A}\,\overline{B}\,C\overline{D}$ coordinates meet, in the upper right square. In a like manner, plot each of the remaining six minterms, in each case finding the one square that satisfies every variable condition in the minterm.

LOOPING MINTERMS TOGETHER

Looping, the technique of grouping minterms together when they fall in adjacent squares, is the way you simplify terms with a Karnaugh map. If no minterms occupy adjacent squares, the expression you mapped can't be minimized: it's already in its simplest form.

Loops can be made only around groups of two, four, eight, or sixteen adjacent minterm positions—all powers of 2. Also, the squares holding the minterms being looped must be adjacent horizontally or vertically, not diagonally.

The map plotted in **Figure 6-19c** immediately reveals one possible group of four: 2, 3, 7, and 6. They form a vertical line of adjacent squares. There's another loop of two minterms—4 and 8.

Always loop as many terms as possible, becuase that minimizes terms the farthest. However, you can't include 4 and 8 in the 2-3-7-6 loop, because—remember?—you can't have six minterms in a loop.

However, a different loop of four is possible: minterms 3, 4, 8 and 7. For reasons that will soon be apparent, you'd loop these together, instead of the others mentioned. This has been done in **Figure 6-19c**.

"ROLLING" A KARNAUGH MAP

The top edge of a Karnaugh grid is considered adjacent to the bottom edge, as if

Digital Circuits and Devices

the map were rolled into a cylinder. As **Figure 6-20a** shows, this places all the squares along the top edge adjacent to all of those along the bottom.

Consequently, minterms 5 and 1 are adjacent in the map, and so are 6 and 2. You could therefore form two more loops around these pairs.

Left and right edges are also adjacent. **Figures 6-20b** indicates the result of this "rolling": minterms 2 and 1 now prove to be adjacent, as do 6 and 5.

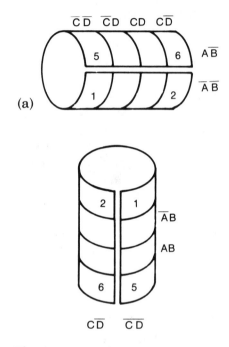

Figure 6-20. "Rolling" a Karnaugh map

In figuring out the loops in a Karnaugh map, learn to exploit these various adjacencies. For this purpose, the map can be considered as rolled both ways at once. So, the four corners of the map in **Figure 6-19c** form adjacent squares, too. Hence, the minterms 1, 2, 6, and 5 fit into a loop of four. The type of loop that encompasses them has already been drawn for you in **Figure 6-19c**.

Thus it can be seen that two loops embrace all of the eight minterms of the original equation. These loops each form a new minterm, provided you know how to "read" them.

READING KARNAUGH LOOPS

To read a loop, trace from square to square inside it, moving either horizontally or

6
Digital Simplification

vertically. You'll discover that each square inside a loop contains one variable that changes state as you move to the adjacent square. Being complementary, these variables cancel.

For example, work your way around the 3-4-8-7 loop in **Figure 6-19c**. From 3 to 4, D cancels D-not, leaving C. From 4 to 8, A cancels A-not, leaving B. From 8 to 7 is the same as from 3 to 4, and from 7 to 3 is the same as from 4 to 8. Consequently, this loop evaluates to a logical result of BC. This *loop value* thus forms one minterm of a new, equivalent Boolean expression.

Turning to the 1-2-6-5 loop, you find C and C-not canceling from 1 to 2, and from 6 to 5. What's left is D-not. In the move from 2 to 6, and also from 5 to 1, A cancels A-not, leaving only B-not. This loop thus evaluates to $\overline{B}\overline{D}$, another minterm in the new equation.

Since two loops have encircled all the original minterms, the new expression has only two minterms. The equation, minimized, has become

$$X = BC + \overline{B}\overline{D}$$

So, while it may be hard to believe, the complex eight-term, four-input equation you began with has reduced logically to two minterms and only three inputs.

FROM EQUATION TO LOGIC CIRCUITRY

What you've accomplished is a simplification of the circuitry for our numeric decoder. The equivalent Boolean expression quoted above means that X proves true either when B and C are both *high* or when B and D are both *low*.

Two questions arise: Does this truly satisfy the decoder requirement, and what happened to A, if this is a four-input decoder?

To answer the first, draw the minimized logic circuit, which should look like **Figure 6-21a**. That is, it takes two AND gates, two NOT gates, and an OR gate. The diagram should agree exactly with the new, shorter Boolean expression.

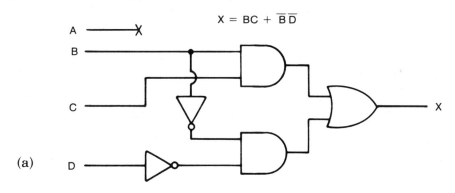

Figure 6-21. Minimized logic circuitry with truth table

151

Digital Circuits and Devices

A	B	C	D	X	Decimal
0	0	0	0	1	0
0	0	0	1	0	1
0	0	1	0	1	2
0	0	1	1	0	3
0	1	0	0	0	4
0	1	0	1	0	5
0	1	1	0	1	6
0	1	1	1	1	7
1	0	0	0	1	8
1	0	0	1	0	9
1	0	1	0	1	10
1	0	1	1	0	11
1	1	0	0	0	12
1	1	0	1	0	13
1	1	1	0	1	14
1	1	1	1	1	15

(b)   **Figure 6-21.** (continued)

Does this circuit decode only those numbers desired, and no others? To help you see, a chart of possible input patterns is offered in **Figure 6-21b**. For your own enlightenment, verify each line of the truth table. Trace the output logic for every binary pattern at inputs B, C, and D.

Note that the three least significant bits of the binary patterns are the same for decimal 0 and 8, for 2 and 10, for 6 and 14, and for 7 and 15. Therefore, the decoder delivers logic 1 to output X whenever these combinations appear at inputs B, C, D. And because the decoder turns on whether line A is logic 1 or logic 0, line A need not even be connected. The Karnaugh simplification procedure "discovered" this as it minimized the logic necessary to accomplish this function, and, to answer the second question, it eliminated A from the Boolean equation.

Topic Review   6-7

1. A Karnaugh map charts ANDed variables called _____.

2. A _____ _____ is a new minterm resulting from Karnaugh simplification.

3. _____, the technique of grouping minterms together when they fall in adjacent squares, is the way to simplify terms with a Karnaugh map.

Answers:

1. minterms
2. loop value
3. Looping

6-8 Summary Points

1. Boolean algebra is a system of theorems and postulates used in the design of digital electronic circuitry.

2. AND, OR, and NOT are fundamental logic operators. All digital circuitry uses some combination of these basic functions.

3. AND identities show the relationship between the inputs of an AND gate and the output condition of the AND gate.

4. OR identities show the relationship between the inputs of the OR gate and the output condition of the OR gate.

5. NOT identities show the relationship between the input and output of an inverter.

6. Boolean laws allow the user to rearrange a Boolean expression without changing its function.

7. The five Boolean laws are the associative, commutative, associative and commutative, distributive, and DeMorgan's laws.

Digital Circuits and Devices

8. Boolean theorems are rules used to reduce Boolean expressions to their lowest form.

9. A Karnaugh map is a method used to find the most compact logical equivalent of a complex truth table or Boolean expression.

10. A minterm is a common form of a logic equation derived from ANDed variables used in simplifying digital-logic variables by Karnaugh mapping.

6-9 Chapter Progress Evaluation

Simplify the following Boolean expressions to their lowest terms:

1. [AB + [[DC] + [$\overline{AD}$] + [$\overline{A B C}$]]]

2. [[[A + B + C] $\overline{AB}$] + [$\overline{A + B}$]]

3. [[$\overline{A + B}$] + C]

4. [[[$\overline{A + B}$] + D] · [[$\overline{C B}$] $\overline{D}$]]

5. [[[$\overline{A B C D}$] + [$\overline{A B}$]] + [[[CD][AD]] + B]]

Using Karnaugh mapping simplify the following Boolean expressions to their lowest form:

6. [[$\overline{A}$BC] + [$\overline{B}\,\overline{C}$] + C + $\overline{D}$]

7. [[$\overline{A}\,\overline{B}\,\overline{C}$D]+[A$\overline{B}\,\overline{C}$]+ D +[$\overline{A}$B]]

8. [[$\overline{A}\,\overline{B}$C]+[$\overline{A}\,\overline{B}$]]

9. [[$\overline{B}\,\overline{C}$ D] + [$\overline{A}\,\overline{B}$] + [AB$\overline{C}$ D]]

10. [[A $\overline{B}$] + B + C + $\overline{C}$]

Chapter 7

Digital Integration

Objectives

Upon completion of this chapter, you should be able to do the following:

- Describe the purpose and operation of a half adder, full adder, parallel adder, half subtractor, and full subtractor

- Describe the operation of a multiplexer, demultiplexer, decoder and encoder

- List and describe the operation of common digital displays

- Describe the operation of various display decoder/drivers

- Describe the purpose of expandable gate units, Schmitt trigger gates, bilateral switches, magnitude comparators, parity generators, and parity checkers

7-1 Introduction

Digital integration allows many basic logic gates to be integrated onto a single chip, forming a complex digital function. In the area of performing mathematical operations, some of the digital functions available are half adders, full adders, half subtractors, full subtractors, and magnitude comparators. Other digital functions include multiplexers, demultiplexers, decoders, encoders, digital displays, digital drivers, bilateral switches, parity generators, and parity checkers. All of these functions resulting from digital integration will be studied in this chapter.

ARITHMETIC OPERATIONS

Binary Addition and Subtraction

Before studying some digital circuits that are used to perform arithmetic operations, let's first look at the rules for binary addition and subtraction.

Digital Circuits and Devices

The rules for *binary addition* are:

    Rule 1 — $0 + 0 = 0$

    Rule 2 — $0 + 1 = 1$

    Rule 3 — $1 + 0 = 1$

    Rule 4 — $1 + 1 = 0$ *and carry* 1

    Rule 5 — $1 + 1 + 1 = 1$ *and carry* 1

The rules for *binary subtraction* are:

    Rule 1 — $0 - 0 = 0$

    Rule 2 — $0 - 1 = 1$ with a borrow

    Rule 3 — $1 - 0 = 1$

    Rule 4 — $1 - 1 = 0$

    Rule 5 — $10 - 1 = 1$

Parallel and Serial Arithmetic

Two kinds of logic devices perform arithmetic functions in digital systems. One is the exclusive-OR gate, and the other is a shift register. In one combination or another, these fundamental devices can do virtually every arithmetic manipulation of binary data.

Addition, subtraction, multiplication, and division can be accomplished through one form or another of binary addition. Hence the key digital stage in arithmetic units is known as an *adder*.

Data, in the form of bit groups, can be fed to adder registers serially, one bit at a time. A *serial* adder unit requires shift registers, a data input line for each register, a clock line, and an output register and line.

A *parallel* adder unit can do the same job in far less time. A parallel input bus delivers whole nibbles or bytes of data into the arithmetic registers at one time. However, a parallel adder requires more circuit wiring and a synchronous clocking scheme.

Basic Half Adders

A *half adder* has one limitation in that it can't handle a *carry-in* which is a carry from some previous binary arithmetic operation. Despite its limitation of not being able to handle a carry-in, a half adder can produce a *carry-out*. That is, if combining two digital bits results in a carry, the half adder can so indicate to whatever stage follows.

Figure 7-1a diagrams a simple half adder. It consists of an exclusive-OR gate and one AND gate to conduct the carry, if there should be one. The circuit is usually shown on logic diagrams in its combined form as illustrated in **Figure 7-1b**.

A and B represent input lines for the bits to be added. S is the sum output. C is the carry output. This single stage is a one-bit adder, capable of adding two one-bit binary numbers.

Digital Integration

The truth table in **Figure 7-1c** explains the operation of the half adder. Logic 0 at both A and B — adding 0 + 0 — leaves the ex-OR and AND gate outputs *low*. Logic 0 appears at both outputs, S and C.

Logic 1 at B only turns on the ex-OR gate, but not the AND gate. The output at S is logic 1, and at C is logic 0 — the sum of 0 + 1. Logic 1 at A only, produces the same output at S and C.

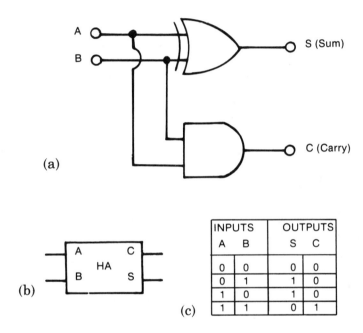

INPUTS		OUTPUTS	
A	B	S	C
0	0	0	0
0	1	1	0
1	0	1	0
1	1	0	1

Figure 7-1. The half adder

Logic 1 at both A and B, representing 1 + 1, turns on the AND gate, but leaves the ex-OR output *low*. Therefore, the output at S is logic 0, and the output at C is logic 1. Adding 1 + 1 produces a sum of 0 and a carry-out of 1 from the adder.

Notice that the truth table for the half adder corresponds exactly with the first four of the five rules for binary addition. Nevertheless, half adders are seldom used in IC adder units, since their inability to handle a carry-in prevents cascading.

Full Adders

Essentially, a *full adder* sums three inputs, one of which transfers a carry-in from some previous arithmetic function. The logic diagram in **Figure 7-2a** shows the use of two half adders: one for the addend and augend, and another for the

Digital Circuits and Devices

carry-in. An *addend* is the number or quantity to be added to another quantity called the *augend*.

HA-1 performs just like any half adder, summing the A and B inputs. If a carry is produced, it goes to the OR gate, signaling it to send a carry-out. Meanwhile, the sum goes to HA-2, to be added with the C input. If this binary addition produces a carry, it also goes to the OR gate. Thus, a carry generated by either half adder appears at the C output of the full adder. The sum produced at S of HA-2 becomes the sum output of the full adder.

The truth table in **Figure 7-2c** verifies every input combination of these three inputs. The bottom line of this truth table corresponds to the fifth rule for binary addition. Notice that the full adder is a logic circuit which performs the five rules of binary addition.

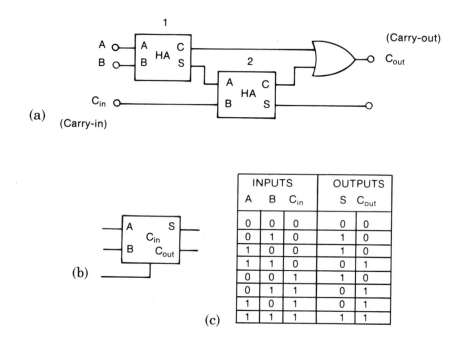

Figure 7-2. The full adder

The full adder of **Figure 7-2** is a one-bit adder. Practical adders are larger, usually handling four and eight-bit numbers. A multibit adder could consist of a half adder for bit 0 (which wouldn't have to accomodate a carry-in), and enough other full adders to manipulate the other bits.

7
Digital Integration

Figure 7-3 illustrates a four-bit full adder, which is also sometimes called a four-bit *parallel adder*. Notice that in addition to accepting a four-bit augend and a four-bit addend, this cascaded adder has a C_{in} terminal and a C output labeled C_4. Multi-bit adders are built in this manner with all full adders cascaded end-to-end in order to handle the addition of large binary numbers. Notice that the C terminals transfer carry bits from one full adder to the next.

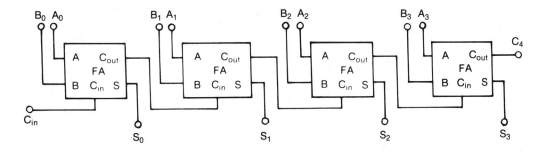

Figure 7-3. A four-bit adder

Logic Subtractors

Figure 7-4a shows the digital gates that make up a *half subtractor* along with its corresponding truth table. Notice that this digital circuit resembles the half adder with the exception of the inverter gate. The result of using this inverter is a *difference* rather than a sum, and a *borrow* rather than a carry. The truth table for the half subtractor shows that it can perform the first four of the five rules for binary subtraction.

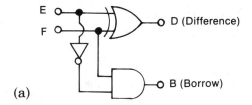

INPUTS		OUTPUTS	
E	F	D	B_{out}
0	0	0	0
0	1	1	1
1	0	1	0
1	1	0	0

(a)

Figure 7-4. Logic diagrams for the half and full subtractors along with their corresponding truth

159

Digital Circuits and Devices

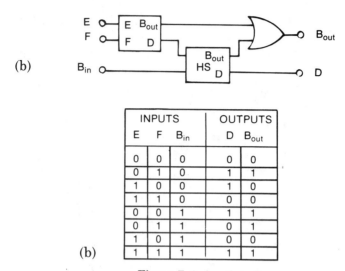

(b)

Figure 7-4. (continued)

A *full subtractor* is diagrammed in **Figure 7-4b**, along with its truth table. As you might expect, a full subtractor accommodates a borrow from an adjacent bit position. Subtractors can be cascaded to allow multibit subtraction.

Topic Review   7-1

1. A _____ is a carry from some previous binary arithmetic operation.

2. The limitation of a half adder is that it cannot handle a _____.

3. A _____ _____ is the smallest type of adder required to add three single bits.

4. A multibit adder is obtained by _____ _____ _____.

5. A _____ _____ can perform the five rules of binary subtraction.

Answers:

1. carry-in
2. carry-in
3. full adder
4. cascading full adders
5. full subtractor

7
Digital Integration

7-2 Multiplexers

A *multiplexer* is a digital device with many inputs and one output. A multiplexer acts as a unidirectional electronic rotary switch. The purpose of a multiplexer is to allow only the data from one of its inputs to be displayed on its output at any one time. A multiplexer is sometimes referred to as a *data selector*.

Besides the inputs and the output, multiplexers also contain *select* (control) lines and usually *strobe* (enable) lines. The purpose of the select lines is to select the data from which input will be displayed on the output of the circuit. The number of select lines needed to select one of the inputs is always a power of two. Therefore, one select line can select between two different inputs; two select lines can select between four different inputs and so on. For each select line added, the number of inputs the multiplexer can select from doubles. The purpose of the strobe (enable) line/lines is to allow the multiplexer to work normally or to disable the multiplexer. (Disabling the multiplexer means the output of the circuit will not reflect any of the input data even though one of the inputs are selected.) It should also be noted that the number of inputs a multiplexer has is always a power of two. The most common values are 2, 4, 8 and 16. The 16 input multiplexer is normally the limit due to the number of pins required for this IC multiplexer function.

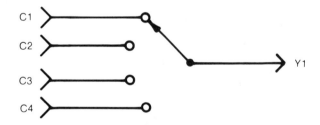

Figure 7-5. Switch representation of a 4 to 1 multiplexer

Figure 7-5 shows a switch representation of a 4 to 1 multiplexer. Note that the four inputs are labled C0, C1, C2 and C3. The output is labeled Y1. The labeling of multiplexers will change from multiplexer to multiplexer and from manufacturer to manufacturer; however, there are always more inputs than there are outputs.

Figure 7-6a shows the gate circuit for one-half of a dual 4 to 1 multiplexer/data selector (74153). As the name indicates, there are two 4 to 1 multiplexers in this IC package. In the diagram there are four inputs (1C0, 1C1, 1C2 and 1C3). The number one in front of each letter C represents that these are the inputs for multiplexer one in the package. The number following the letter C represents the decimal value of the binary code needed on the select lines to select that input. The two select lines are labeled A and B where each are inverted to also produce their complements. These inverted lines are used along with the original non-inverted A and B lines. They are

Digital Circuits and Devices

then applied to each AND input in a different order to allow only one AND gate to be selected at any one time. There is also a strobe line labeled G1 which is inverted before being applied to each AND gate of the input control circuit. When a *high* is placed on the G1 input, a *low* will be applied to each AND gate. This will cause each AND gate to produce a *low* at its output. No matter what the other inputs see, the output of the multiplexer will always be *low*. This is called the *disabled* state. When a *low* is placed on the G1 input, the multiplexer will operate in its normal manner. The output of the multiplexer is labeled Y1. The number following the letter Y represents that this is the output from multiplexer one in the package.

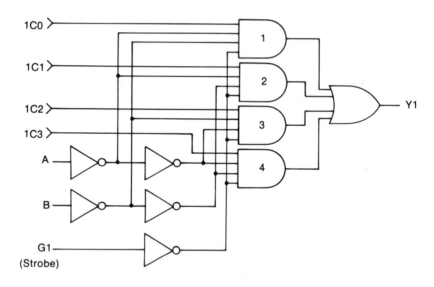

SELECT A B	STROBE G1	AND GATE 1 2 3 4	OUTPUT Y1
L L	L	EN DIS DIS DIS	1C0
L H	L	DIS EN DIS DIS	1C1
H L	L	DIS DIS EN DIS	1C2
H H	L	DIS DIS DIS EN	1C3
X X	H	DIS DIS DIS DIS	L

L = Low logic level   EN = Enable gate
H = High logic level   DIS = Disable gate
X = Does not matter what level

Figure 7-6. Logic diagram and corresponding truth table for a 4 to 1 multiplexer/data selector

Digital Integration

In the 74153 dual 4 to 1 multiplexer/data selector, the select lines are shared by both multiplexers in the package, so both multiplexers will select the same inputs at the same time. In addition there are separate strobe lines for each multiplexer which will allow the user to disable them separately.

From the internal truth table and the logic diagram of **Figure 7-6**, you should note the following gate input connections

    AND gate 1 = 1C0, A NOT, B NOT and G1 NOT
    AND gate 2 = 1C1, A NOT, B and G1 NOT
    AND gate 3 = 1C2, A, B NOT and G1 NOT
    AND gate 4 = 1C3, A, B and G1 NOT
    OR gate   = AND 1, AND 2, AND 3 and AND 4

When select lines A and B are *low* and the strobe is *low*, the data applied to input 1C0 will be applied to the OR gate. All other AND gates at this time will have a *low* on their outputs which will not affect the OR gate of the circuit. The output of the multiplexer at this time will be equal to the data applied to input.

When select line A is *low* and B is *high* and the strobe is *low*, the data applied to input 1C1 will be applied to the OR gate. All other AND gates at this time will have a *low* on their outputs which will not affect the OR gate of the circuit. The output of the multiplexer at this time will be equal to the data on 1C1.

When select line A is *high* and B is *low* and the strobe is *low*, the data applied to input 1C2 will be applied to the OR gate. All other AND gates at this time will have a *low* on their outputs which will not affect the OR gate of the circuit. The output of the multiplexer at this time will be equal to the data on 1C2.

When select lines A and B are both *high* and the strobe is *low*, the data applied to input 1C3 will be applied to the OR gate. All other AND gates at this time will have a *low* on their outputs which will not affect the OR gate of the circuit. The output of the multiplexer at this time will be equal to the data on 1C3.

Whenever the strobe input (G1) is *high*, a *low* will be placed on the input to each AND gate, which will produce a low at the output of each AND gate. These *low* logic levels will never change unless the strobe is brought *low* again. Since all the outputs of the AND gates are *low*, the output from the OR gate of the circuit will be *low*. This means none of the data from the inputs is seen by the output. When this condition occurs, the multiplexer is said to be in the disabled state.

Multiplexers/data selectors serve many functions in digital electronics. They can be used to convert parallel binary data to serial binary data. This allows communications between two digital devices over a single pair of wires rather than needing a pair of wires for each binary bit. Multiplexers can be connected to various devices, and any one of these devices from which it wishes to take data.

Digital Circuits and Devices

The following are common TTL multiplexers:

74150 16 to 1 multiplexer/data selector
74151 8 to 1 multiplexer/data selector
74152 8 to 1 data selector
74153 dual 4 to 1 multiplexer/data selector
74157 quad 2 to 1 data selector/multiplexer (non-inverting)
74158 quad 2 to 1 data selector/multiplexer (inverting)

7-3 Demultiplexers

A *demultiplexer* is a digital device with one input and many outputs. A demultiplexer acts as a unidirectional electronic rotary switch. The purpose of a demultiplexer is to allow data from the one input of the circuit to be applied to any one of the many outputs of the circuit. Demultiplexers are often called *decoders*. The major difference between the two is that demultiplexers allow changing data from its input to be applied to one of its outputs. In a decoder, the data bits are fixed. Being applied to the input line/lines of a decoder, these data bit/bits select the output line that will be activated.

Besides the input and outputs, demultiplexers also contain select (control) lines and usually strobe (enable) lines. The purpose of the select lines is to select which output will see the data from the input. The number of select lines needed to select one of the outputs is always a power of two. Therefore, one select line can select between two different outputs; two select lines can select between four different outputs and so on. For each select line added, the number of outputs the demultiplexer can select from doubles. The purpose of the strobe (enable) line/lines is to allow the demultiplexer to work in its normal fashion or to disable the demuiltiplexer meaning all the outputs will be disabled. When this occurs, the selected output will not see the input data. It should also be noted that in some demultiplexer/decoder chips, the enable lines are used as the data input. The number of outputs a demultiplexer has is also a power of two. The most common values are 2, 4, 8 and 16. The 16 output demultiplexer is normally the limit due to the number of pins required for the demultiplexer function.

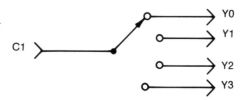

Figure 7-7. Switch representation of a 1 to 4 demultiplexer

7
Digital Integration

The diagram of **Figure 7-7** shows a switch representation of a 1 to 4 demultiplexer. Note there is one input called C1, and four outputs labeled Y0, Y1, Y2 and Y3. The labeling of demultiplexers will change from demulitplexer to demultiplexer and from manufacturer to manufacturer; however, there are always more outputs than there are inputs.

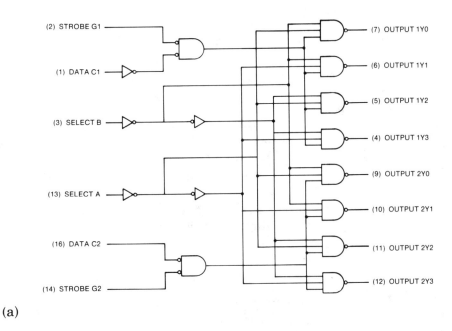

(a)

2-LINE-TO-4—LINE DECODER
OR 1-LINE-TO-4-LINE DEMULTIPLEXER

INPUTS				OUTPUTS			
SELECT		STROBE	DATA	1Y0	1Y1	1Y2	1Y3
B	A	G1	C1				
X	X	H	X	H	H	H	H
L	L	L	H	L	H	H	H
L	H	L	H	H	L	H	H
H	L	L	H	H	H	L	H
H	H	L	H	H	H	H	L
X	X	X	L	H	H	H	H

INPUTS				OUTPUTS			
SELECT		STROBE	DATA	2Y0	2Y1	2Y2	2Y3
B	A	G2	C2				
X	X	H	X	H	H	H	H
L	L	L	L	L	H	H	H
L	H	L	L	H	L	H	H
H	L	L	L	H	H	L	H
H	H	L	L	H	H	H	L
X	X	X	H	H	H	H	H

(b)

Figure 7-8. Logic diagram and corresponding truth tables for the 74156 demultiplexer/decoder

165

Digital Circuits and Devices

Figure 7-8a shows a logic diagram for a dual 4 to 1 demultiplexer/decoder (74156). As the word dual indicates, there are two 1 to 4 demultiplexers in this IC package. In the logic diagram there are four outputs for each demultiplexer. Outputs 1Y0, 1Y1, 1Y2 and 1Y3 are the outputs for demulitplexer one, while outputs 2Y0, 2Y1, 2Y2 and 2Y3 are the outputs for demultiplexer two. The select lines A and B are common for each demultiplexer, so each demultiplexer will select the same output to display their data. There are two separate data inputs. Data input C1 for demultiplexer one is inverted, so the data on the selected output will always be the opposite of the data on the input. Data input C2 for demultiplexer two is a non-inverted data input, so the data on the selected output will be the same as the data on the input. There are also two strobes labeled G1 and G2 where both are active *lows*.

In other words when the strobes are *low*, the demultiplexer will operate normally, whereas when they are *high*, the outputs will be disabled. In this case when the demultiplexer is disabled, the output will be *high* due to the fact that the output control circuitry is derived from NAND gates.

The operation of the 74156 demultiplexer/decoder can be derived from the truth tables shown in **Figure 7-8b**.

Demultiplexers serve many functions in digital electronics. They can be used to convert serial binary data into parallel binary data. Demultiplexers can select one of many different digital devices to talk to.

The following are common TTL demultiplexers:

    74138 1 to 8 demultiplexer/decoder
    74139 dual 1 to 4 demultiplexer/decoder
    74154 1 to 16 demultiplexer/decoder
    74156 dual 1 to 4 demultiplexer/decoder
    74159 1 to 16 demultiplexer/decoder

7-4 Fundamentals of Encoding and Decoding

Binary representations of numeric quantities are generally referred to as *codes*. In this context, the word *encoding* means converting a quantity to its binary equivalent.

However, in a broader digital logic context, the meaning can go deeper. For example, activating certain segments of the seven segments of an LED (light-emitting diode) display device forms visible letters or numerals. The logic pattern that lights them is a binary code. Creating the proper pattern of bits to display a numeral 2, for example, is encoding.

Decoding, on the other hand, can be thought of as *interpreting* a binary code or pattern. The logic in a decoder is wired to recognize a particular arrangement of bits,

7
Digital Integration

and then to produce a certain output. One common use of a decoder is to recognize a unique binary *address* and use it to enable a memory chip as in a computer or programmable controller. In such an application, a decoder may also be called a *detector*.

Sometimes a decoder output re-encodes the data. A typical decoder for the LED display mentioned above does just that. A BCD-to-seven-segment decoder accepts a four-bit BCD pattern which is the binary representation of a decimal digit. Upon recognizing or *decoding* a valid BCD code, the device proceeds to generate a new binary code containing seven bits. This is encoding. The new pattern, when applied to a seven-segment LED, displays the decimal digit indicated by the original BCD code.

Digital *multiplexing* and *demultiplexing* bear a close resemblance to encoding and decoding. In fact, decoders are often called *demultiplexers*. However, multiplexing is slightly different from the concept of encoding.

1-of-8 DECODERS

Sequential binary decoders have the ability to select one output from among several. The output line selected depends on a binary pattern fed to the control or selector inputs.

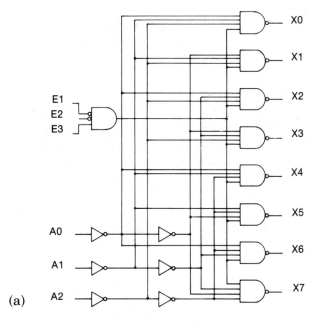

(a)

Figure 7-9. Logic diagram and corresponding truth table for a 1-of-8 decoder

Digital Circuits and Devices

ENABLE LINES			ADDRESS LINES			OUTPUT LINES							
E3	E2	E1	A2	A1	A0	X7	X6	X5	X4	X3	X2	X1	X0
0	0	0	X	X	X	1	1	1	1	1	1	1	1
0	0	1	X	X	X	1	1	1	1	1	1	1	1
0	1	0	X	X	X	1	1	1	1	1	1	1	1
0	1	1	X	X	X	1	1	1	1	1	1	1	1
1	0	0	0	0	0	1	1	1	1	1	1	1	0
1	0	0	0	0	1	1	1	1	1	1	1	0	1
1	0	0	0	1	0	1	1	1	1	1	0	1	1
1	0	0	0	1	1	1	1	1	1	0	1	1	1
1	0	0	1	0	0	1	1	1	0	1	1	1	1
1	0	0	1	0	1	1	1	0	1	1	1	1	1
1	0	0	1	1	1	1	0	1	1	1	1	1	1
1	0	0	1	1	1	0	1	1	1	1	1	1	1
1	0	1	X	X	X	1	1	1	1	1	1	1	1
1	1	0	X	X	X	1	1	1	1	1	1	1	1
1	1	1	X	X	X	1	1	1	1	1	1	1	1

(b)

Figure 7-9. (continued)

Consider the decoder in the logic diagram of **Figure 7-9a**. A three-input AND gate allows the *triple enabling* of the decoder for the purpose of cascading several such devices. For purposes of this explanation, assume that this enabling gate receives logic that turns it on.

Inputs A2, A1 and A0 accept the binary pattern to be decoded. If the pattern is 000, the NAND gate at X0 receives enabling logic for all three remaining inputs. Enable logic keeps the fourth input *high*. Therefore, output X0 is the only line pulled to logic 0. All other output lines remain logic 1.

Applying the other binary patterns to the address lines (A2, A1, and A0) and analyzing the gates of the logic diagram results in the truth table of **Figure 7-9b**. In every case the decoder has selected one output from among the eight possible, pulling that output terminal to logic 0. Hence the designation *1-of-8 decoder*. Also notice that the decoder is enabled only when the enable lines E3, E2, and E1 are at logic 1, 0, and 0, respectively.

Such a decoder could select one of eight different memory banks, turn on one of eight different process-control valves, or turn on one of eight lamps. In each case only a three-bit *address* is needed to select one of the eight devices.

The decoder can operate in another way. Suppose that the E1 and E2 inputs were tied to logic 0. A binary pattern at the A inputs would select an output gate, but the output from that gate would depend on the logic level at the E3 input. Logic 1 at E3 would drive the selected output line to logic 0. Logic 0 at E3 would return that output to logic 1. Thus, any data stream at E3 would be fed inverted through the selected output gate. All other output gates would remain *high*, not having been selected by

168

7
Digital Integration

the A input pattern.

Changing the address pattern would reroute the data stream entering at E3. This is another function this type of decoder or demultiplexer can perform. Complex decoders can direct the paths of multibit patterns, and a gate circuit made up of a number of such decoders in parallel could control a byte or more at a time.

ENCODERS

An *encoder* converts a quantity into a binary or BCD equivalent. One example of an encoder is the 74147 10-line to 4-line BCD priority encoder. This device has 10 decimal inputs and the output BCD code will be determined by the decimal input. This type of encoder is used to produce a BCD output code when a key on a keyboard is depressed.

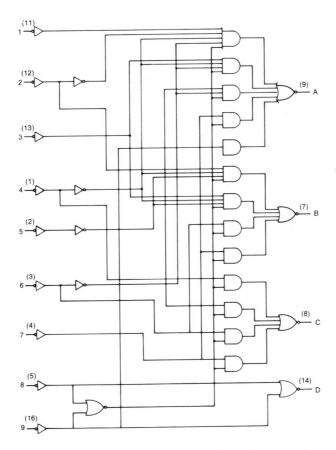

Figure 7-10. The 74147 10-line to 4-line priority encoder

169

Digital Circuits and Devices

INPUTS									OUTPUTS			
1	2	3	4	5	6	7	8	9	D	C	B	A
H	H	H	H	H	H	H	H	H	H	H	H	H
X	X	X	X	X	X	X	X	L	L	H	H	L
X	X	X	X	X	X	X	L	H	L	H	H	H
X	X	X	X	X	X	L	H	H	H	L	L	L
X	X	X	X	X	L	H	H	H	H	L	L	H
X	X	X	X	L	H	H	H	H	H	L	H	L
X	X	X	L	H	H	H	H	H	H	L	H	H
X	X	L	H	H	H	H	H	H	H	H	L	L
X	L	H	H	H	H	H	H	H	H	H	L	H
L	H	H	H	H	H	H	H	H	H	H	H	L

H = High Logic Level   L = Low Logic Level   X = Don't Care

Figure 7-10. (continued)

Topic Review   7-2 through 7-4

1. A _____ is an electronic device that will select one of its many inputs to be displayed on its output.

2. _____ _____ are used to control which output of a demultiplexer will display the input data.

3. The purpose of a _____ _____ for multiplexers and demultiplexers is to provide a means of disabling these devices.

4. A _____ converts its input data in the form of a binary code to select one of its output lines.

5. The term _____ refers to converting a quantity to its binary equivalent.

Answers:

1. multiplexer
2. select lines
3. strobe line
4. decoder
5. encoding

7
Digital Integration

7-5 Digital Displays

A digital display allows a visual indication of digital data or codes. Some types of displays create their own light source, and some displays use environmental light for the visual indication.

LIGHT GENERATING DISPLAY SOURCES
Gas Discharge Tubes

One type of light generating digital display is called a *gas discharge tube*. The gas discharge tube is a small glass container with electrodes on both ends. Inside the glass container, one electrode acts as the cathode and the other acts as the anode. When the potential applied between the cathode and anode is high enough, the gas inside the tube will ionize creating light. The type of gas inside the tube will determines the color of the light.

The gas discharge tube was the first type of digital display. One of the problems with it is the glass tube tends to leak causing air to get into the tube. This will reduce the amount of light generated and may even cause the tube not to light at all. Since the tubes are made of glass, they are not very resistant to physical shocks and vibration. The last problem is the voltage necessary to cause the gas to ionize is between 48 and 180 V. This requires the digital levels to be boosted to a high voltage by a separate electronic circuit. The one advantage this device does have is the displays are easy to see in bright light. The only place you will find this type of display source is in old equipment since it is very seldom used in the electronic designs of today.

Light Emitting Diodes

The *light emitting diode* (LED) is a specifically doped multilayer diode using gallium arsenide and gallium arsenide phosphide instead of silicon to produce light when the diode is forward biased. The light is created by the combination of electrons and holes in the *PN* junction which creates photons. The *PN* junction is covered with a lens to condense the light, and a color filter is used which will allow only one color of light to pass through it. The normal color and the easiest to see is red, but other colors are available.

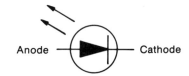

Figure 7-11. Symbol for the LED

Digital Circuits and Devices

The symbol for an LED is shown in **Figure 7-11**. The two arrows exiting the symbol represent light being emitted from the device. The voltage drop needed to forward bias most LEDs is 1.3VDC at a current of 5mA. In other words, to properly forward bias most LEDs, the anode must be about 1.3V higher than the cathode when 5mA of current is flowing through the device. These values do change somewhat from LED to LED.

The main advantages of using LEDs as light generating display sources are low cost coupled with the ability to withstand vibration and external shocks. The major disadvantage of LEDs is that a relatively high current is needed for operation.

NON-LIGHT GENERATING DISPLAY SOURCES

Liquid Crystal Display

The latest type of digital display source is called a *liquid crystal display* (LCD). The LCD uses the reflection of light and the absence of light to display the information.

Each LCD segment cell contains a *nematic gell*. This nematic gell consists of liquid crystal material that is sandwiched between two glass plates with a polarizing filter on top of one glass plate and a reflective piece of glass placed over the other glass plate.

The liquid crystal material (nematic gell) is placed into an etched cavity to form each segment. The nematic gell is made up of little rod shaped semi-conductor materials that are perpendicular to the polarized filter when no voltage potential is applied to the nematic gell. This behavior of the liquid crystal material is illustrated in **Figure 7-12a**.

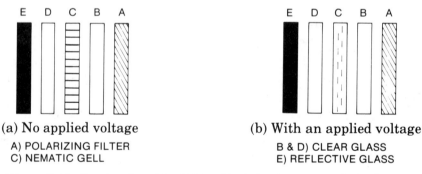

(a) No applied voltage
A) POLARIZING FILTER
C) NEMATIC GELL

(b) With an applied voltage
B & D) CLEAR GLASS
E) REFLECTIVE GLASS

Figure 7-12. Construction of the LCD and its behavior to an applied voltage

Since the crystal material is perpendicular to the filter when no potential is across the nematic gell, light is allowed to pass through the nematic gell. This light is reflected back causing a light gray color to appear on the front of the display. Since the background is also a light gray color, the segment looks like it is off. When a potential is placed across the nematic gell, the rod-shaped semi-conductor material is forced into a parallel angle to the filter as shown in **Figure 7-12b**. This allows no light to be reflected back to the front of the display. The segment will become dark (black), and since the background is light gray, the segment looks like it is on. When the voltage is removed from the nematic gell, the pressure of the gell causes the rods to become perpendicular to the filter which will make the segment look like it is off.

The voltage necessary to operate an LCD display is 16 to 100 V. This allows open collector TTL to drive the displays. Also, since the display is a voltage operated device, the current required for the display is 1 to 2 microamps (μA). The advantages of an LCD display are extremely low power dissipation, and as environmental light becomes brighter, the display is easier to see. The disadvantages are higher cost and the need for some type of external light source in low light environments.

A new type of LCD display used for computers has its own back light source, instead of a reflective piece of glass, on the back of the display to produce the necessary light. This allows the LCD display to be seen in low light areas.

DISPLAY FORMATS

Of the three types of display sources discussed, the gas discharge display normally comes in one format, the 7-segment display. The LED and LCD sources come in the following types of display formats:

Single 7-segment display
Multiple 7-segment display
Dot matrix display
Multiple dot matrix display

The *single 7-segment display* is a device containing seven segments placed in a pattern capable of displaying all numerical characters. The segments are identified as A, B, C, D, E, F, and G as illustrated in **Figure 7-13**.

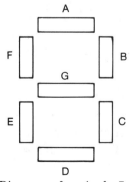

Figure 7-13. Diagram of a single 7-segment display

Digital Circuits and Devices

In each type of display, one end of each segment is tied together to a common point, which may be connected to ground or supply depending on the type of display source. There are also seven segment lines in the display with one connection for each segment. When the proper voltage level is placed on the segment line, the segment will turn on.

The gas discharge, LED, and LCD single 7-segment displays come in either a common cathode or common anode format. In the common cathode format, the cathodes for all segments are connected together are placed at ground potential. In order for a segment to turn on, the segment line must go *high* with respect to the common line. In the common anode format, all anodes for each segment are common and must be applied to the supply. To cause a segment to turn on, the segment line must go *low* for each segment. There are no advantages or disadvantages to which type of format used; however, the common cathode format seems to be more popular. Note if a decimal point is included in the 7-segment display, one additional segment line is needed. This additional segment is labeled DP (decimal point).

Multiple 7-Segment Display

A *multiple 7-segment display* normally contains between two to as many as sixteen single 7-segment displays all in one package. **Figure 7-14** illustrates a package consisting of four 7-segment displays. To cut down on the number of pins required to access this display, all of the seven segment lines are connected together. The common cathode connection for each 7-segment display is kept on separate lines as indicated in the diagram by C1, C2, C3, and C4.

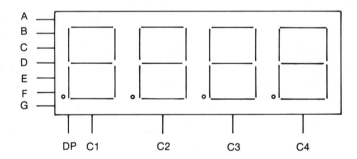

Figure 7-14. Diagram of a multiple 7-segment display

A 7-segment multiplexer/decoder/driver is used to multiplex the 7-segment codes to the multiple display. Here is how it works. When the 7-segment code for the first display is on the segment line, the cathode of the first display (C1) is brought down to ground level. At this time all other cathodes are *high* which means all other 7-segment displays are off. The data will stay on the line for a short time period. Then

7
Digital Integration

the cathode of all displays will go *high* causing all displays to turn off. The 7-segment code for the next display will be applied to the segment lines. The cathode for this display will be applied to the segment lines. The cathode for this display will go *low* turning on the display, while the cathodes to all other displays will be high disabling all other displays. This process continues until all displays in the unit have been accessed, and then starts over again. The process updates 16 to 30 times a second which causes the human eye to think all the 7-segment displays are on continuously.

The advantages that multiple 7-segment displays have over single 7-segment displays are lower cost and less space required. However, a disadvantage is the higher cost of the multiplexer/decoder/driver chip.

Dot Matrix Display

The *dot matrix display* is a device that will display any ASCII character. These characters include letters upper/lower case, symbols, and numbers. Segments are shaped in circular forms and arranged in either a 5 by 7 or 6 by 9, column and row format. The cathodes are usually connected to the rows. In the case of the 5×7 LED dot matrix display shown in **Figure 7-15a**, there are 7 rows with 5 cathodes per row. Connecting the anodes of the LEDs to the columns results in 5 columns with 7 anodes per column.

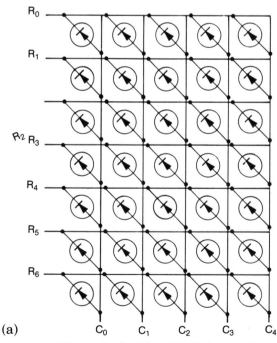

Figure 7-15. A 5×7 LED dot matrix

Digital Circuits and Devices

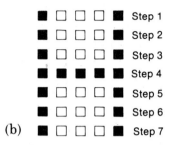

(b)

Figure 7-15. (continued)

In order to gain an understanding of how this 5 × 7 dot matrix works, the following steps outline the process required to cause the letter H to appear on the display as shown in **Figure 7-15b**.

Step 1 Row 0 goes *low*, all other rows will be *high*. Columns 0 and 4 go *high*, all other columns *low*.
10 *msec* time delay.
All columns *low*.

Step 2 Row 1 goes *low*, all other rows will be *high*. Columns 0 and 4 go *high*, all other columns *low*.
10 *msec* time delay.
All columns *low*.

Step 3 Row 2 goes *low*, all other rows will be *high*. Columns 0 and 4 go *high*, all other columns *low*.
10 *msec* time delay.
All columns *low*.

Step 4 Row 3 goes *low*, all other rows will be *high*. Columns 0 through 4 go *high*, all other columns *low*.
10 *msec* time delay.
All columns *low*.

Step 5 Row 4 goes *low*, all other rows will be *high*. Columns 0 and 4 go *high*, all other columns *low*.
10 *msec* time delay.
All columns *low*.

7
Digital Integration

Step 6 Row 5 goes *low*, all other rows will be *high*. Columns 0 and 4 go *high*, all other columns *low*.
10 *msec* time delay.
All columns *low*.

Step 7 Row 6 goes *low*, all other rows will be *high*. Columns 0 and 4 go *high*, all other columns *low*.
10 *msec* time delay.
All columns *low*.

To turn on an LED, the row must go *low* and the column must go *high*. The 10 *msec* time delay allows the display to be seen. When all columns are *low*, the display will blank, allowing the row data to change without seeing the change on the display. This entire process occurs 16 to 100 times per second.

The advantage of a dot matrix display is that any type of character can be displayed. The disadvantage of this display is the cost of both the display and decoder/driver.

Multiple Dot Matrix Display

Dot matrix displays are also available in multiple display formats. This means there are more than one dot matrix display per package. In fact, one of the latest is the 80 character by 25 lines dot matrix display. This type of dot matrix is LCD which offers very low power dissipation. Because of the very complex interfacing between the display and decoder/driver, the decoder/driver is built into the display itself.

Topic Review   7-5

1. The two types of light generating display sources are _____ _____ _____ and _____ _____ _____.

2. The _____ _____ _____ is a non-light generating display source that becomes easier to read as the environmental light becomes brighter.

3. The LCD uses the _____ and _____ of light to display the information.

4. The _____ and _____ display sources come in four basic types of display formats.

5. The _____ _____ _____ is a type of display format that can produce an ASCII character.

Digital Circuits and Devices

Answers:

1. gas discharge tubes, light emitting diodes
2. liquid crystal display
3. reflection, absence
4. LED, LCD
5. dot matrix display

7-6 Display Decoder/Drivers

A *display decoder/driver* or *multiplexer/decoder/driver* will decode the binary, BCD or ASCII codes on the input and produce a 7-segment or dot matrix code to display the proper character on the display. The multiplexer section will multiplex data from one display to another display for multiple display units. The decoder section takes the input code and converts it into the proper segment code to light the correct segments. The driver section supplies enough current and voltage for each segment.

SINGLE 7-SEGMENT DECODER/DRIVER

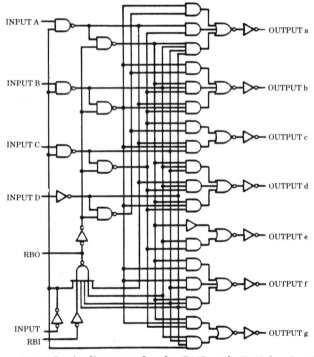

Figure 7-16. Logic diagrams for the 7447 and 7448 decoder/drivers

7
Digital Integration

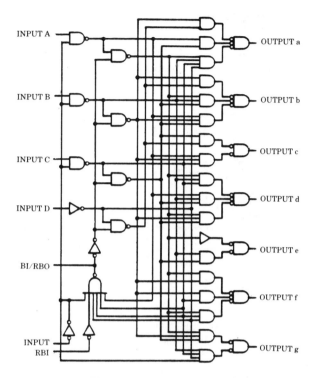

Figure 7-16. (continued)

The two most common display decoder/drivers are the 7447 and 7448. The 7447 is made to work with the common anode 7-segment format display, and the 7448 is made to work with the common cathode 7-segment display. The logic diagrams for the 7447 and the 7448 are shown in **Figure 7-16**.

In the 7447 common anode 7-segment decoder/driver, the anode is connected to the supply, and in order to cause a segment to light up, the segment line must go *low*. The 7447 requires a 200 to 470 ohm current limiting resistor between each segment line and the display itself.

In the 7448 common cathode 7-segment decoder/driver, the cathode is connected to ground, and in order to cause a segment to light up, the segment line must go *high*. The 7448 has current limiting resistors built into the unit.

Besides the input code lines, there are three additional inputs for the 7447 and 7448. The lamp test input (LT) on the 7447 and 7448 will cause all segments of the display to light up when a logic *low* is applied. For normal operation, the LT pin should be tied *high*. The other two input pins on the 7447 and 7448 serve the purpose of supression a leading or lagging zero for the display. RBI (Ripple Blanking In) and

Digital Circuits and Devices

BI/RBO (Blanking In/Ripple Blank Out) are used together to suppress zeros. The application of these two inputs is illustrated in **Figure 7-17**.

The two most left 7448 decoder/drivers are connected to suppress leading zeros and the two most right 7448s are used to suppress any lagging zeros. If suppression is not necessary, these inputs should be tied *high*.

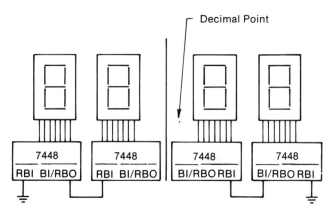

Figure 7-17. Application of the RBI and BI/RBO pins of the 7448 7-segment decoder/driver

MULTIPLEXED DECODER/DRIVER

The *multiplexed decoder/driver* is a device that is used to control and display multiple 7-segment display units. Due to the complex circuitry needed for the controls, this type of decoder/driver is usually produced in CMOS form to reduce size and power dissipation. **Figure 7-18** shows a simple block diagram for a dual display multiplexed decoder/driver.

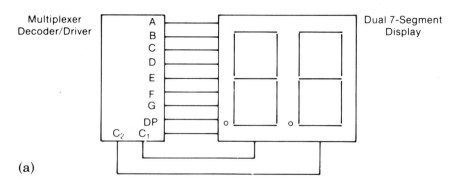

(a)

Figure 7-18. Block diagram of a multiplexed decoder/driver

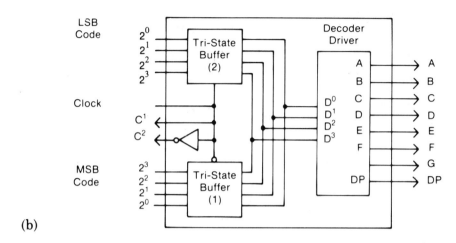

(b)

Figure 7-18. (continued)

Notice that there are nine inputs and eight outputs for the device. The single clock pin controls whether the data from the LSB code or MSB code is applied to the decoder/driver.

The two sets of tri-state buffers have four inputs each and four outputs each. When the clock is *high*, the LSB tri-state buffer is enabled, and the data from the LSB code inputs ($2^0, 2^1, 2^2, 2^3$) will be applied to the decoder/driver. The MSB code tri-state buffer is disabled at this time. This makes tri-state buffer (1) look like it is not part of the circuit. Also at this time, cathode 2 (C2) is *low* which allows the LSB display to turn on. The character on this display will be determined by the code on the segment lines (A, B, C, D, E, F, G, DP). After the display has been on for a short period of time, the clock will go *low*, disabling cathode 2, and the tri-state buffer for the LSB code. Now the tri-state buffer for the MSB code will become enabled, and the data from the MSB code ($2^0, 2^1, 2^2, 2^3$) will be applied to the decoder/driver which will cause a character to be displayed on the MSB display. The process continues back and forth at a rate of 16 to 100 times per second. This will fool the eye making both displays seem like they are on at the same time.

The advantages of a multiplexed decoder/driver display is that it uses less power and requires less board space than separate decoder/driver displays. The main disadvantage is higher cost of the multiplexed decoder/driver IC.

DOT MATRIX DECODER/DRIVER

The *dot matrix decoder/driver* will convert ASCII code and produce a row and column code to allow any ASCII character to be displayed on the dot matrix display.

Digital Circuits and Devices

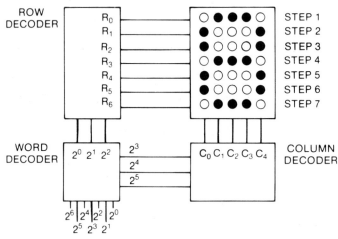

Figure 7-19. Block diagram of a dot matrix decoder/driver being used to drive a single 5 × 7 dot matrix display

Figure 7-19 illustrates a simple block diagram of a 5 × 7 dot matrix decoder/driver. The *word decoder* converts the ACSII code into a row and column code. The internal control in the work decoder, will cause the row and column decoders to enable the proper rows and columns at their proper times; this produces the correct character display.

For example, consider the following steps required to cause the display to show the number eight as shown in **Figure 7-19**.

Step 1 Row 0 goes *low*, all other rows go *high*. Columns 1, 2, 3 go *high*, all other columns go *low*.
10 *msec* time delay.
All columns go *low*.

Step 2 Row 1 goes *low*, all other rows go *high*. Columns 0 and 4 go *high*, all other columns go *low*.
10 *msec* time delay.
All columns go *low*.

Step 3 Row 2 goes *low*, all other rows go *high*. Columns 0 and 4 go *high*, all other columns go *low*.
10 *msec* time delay.
All columns go *low*.

Step 4 Row 3 goes *low*, all other rows go *high*. Columns 1, 2, 3 go *high*, all other columns go *low*.
10 *msec* time delay.
All columns go *low*.

7
Digital Integration

Step 5 Row 4 goes *low*, all other rows go *high*. Columns 0 and 4 go *high*, all other columns go *low*.
10 *msec* time delay.
All columns go *low*.

Step 6 Row 5 goes *low*, all other rows go *high*. Columns 0 and 4 go *high*, all other columns go *low*.
10 *msec* time delay.
All columns go *low*.

Step 7 Row 6 goes *low*, all other rows go *high*. Columns 1, 2, 3 go *high*, all other columns go *low*.
10 *msec* time delay.
All columns go *low*.

After going through these seven steps, the process starts again repeating at a rate of 16 to 100 times per second. This rate depends on the time duration of the various delays.

The advantage of the dot matrix decoder/driver is its ability to display any ASCII character. The main disadvantage is the cost of the unit. Due to the cost and board space required for a separate display and decoder/driver, most applications use this type of display with the decoder/driver built into the unit.

MULTIPLE DOT MATRIX DECODER/DRIVER

The *multiple dot matrix decoder/driver* is a combination of the standard single dot matrix decoder/drivers which are multiplexed together. This type of display can be as large as 80 characters per row and 25 rows per display. It works basically the same as the single dot matrix decoder/driver. Due to the complexity of these devices, they always have their decoder/driver built into the display unit itself.

Topic Review   7-6

1. The _____ and _____ are the two most common 7-segment display decoder/drivers.

2. The _____ is a device that is used to control and display multiple 7-segment display units.

3. The _____ _____ _____ will convert ASCII code and produce a row and column code to allow any ASCII character to be displayed on a _____ _____ _____.

Digital Circuits and Devices

4. ASCII code is converted into a row and column code by a _____ _____.

5. A _____ _____ _____ _____ can be as large as 80 characters per row and 25 rows per display.

Answers:

1. 7447, 7448
2. multiplexed decoder/driver
3. dot matrix decoder/driver, dot matrix display
4. word decoder
5. multiple dot matrix decoder/driver

7-7 Other Digital Integration Functions

EXPANDABLE GATE UNITS

One example of an *expandable gate function* is using the 7450 dual 2-wide, 2-input, AND-OR-INVERTER gate unit and a 7460 dual 4-input expander AND gate. The connection of these two devices is to obtain an expandable gate function as shown in **Figure 7-20a**. The resulting gate function will operate the same as the gate citcuit of **Figure 7-20b**.

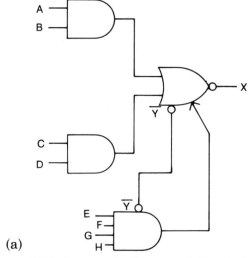

(a)

Figure 7-20. An example of an expandable gate function and its equivalent gate circuit

184

7
Digital Integration

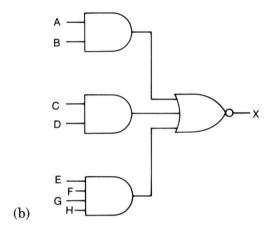

Figure 7-20. (continued)

As can be seen from this equivalent circuit, by connecting the two gate units the number of AND gate inputs can be increased. There are also many other different types of expandable gate functions. Each one allows the user to increase the number of inputs a circuit needs. These types of combination gate units are useful in designing complex gate circuits.

SCHMITT TRIGGER INPUTS

A gate with a *Schmitt trigger* input is used in digital electronics where noise may be introduced on the data line. The Schmitt trigger gate will see a logic 1 if the input voltage is 1.7 V or larger. The output of the gate will not change until the input level goes below 0.9 V. Any noise that may cause the logic *high* to go below the typical 2 V upper threshold level will still be seen as a logic 1 until the voltage on the input decreases below 0.9 V. This process increases the amount of noise margin from 1 V on a typical TTL gate to about 2.1 V. The Schmitt trigger also eliminates the unstable state between the upper threshold and lower threshold input levels.

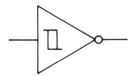

Figure 7-21. The logic symbol for an inverter with a Schmitt trigger input

185

Digital Circuits and Devices

One type of Schmitt trigger gate is the 7414 hex Schmitt trigger inverter. This device contains six inverters with Schmitt trigger inputs. **Figure 7-21** shows the logic symbol of a Schmitt trigger input inverter. The symbol inside the inverter symbol indicates a Schmitt trigger input.

DIGITAL BILATERAL SWITCHES

The 4016 and the 4066 are quad *bilateral switches* which can be used for multiplexing and demultiplexing digital data. The word bilateral means that data can flow in both directions. Each input can act as an input or output and each switch acts as an SPST switch. Each switch also contains its own control line. When the control line is *high*, the switch is closed providing an on-resistance of about 80 ohms for the 4066 and about 130 ohms for the 4016. When the control line is *low*, the switch will be open, and no data will be allowed to pass through the switch. Both the 4016 and the 4066 share the same pinouts as shown in **Figure 7-22**. The only difference is the on-resistance. Also, since these two devices are CMOS, the supply can range from 3 to 15V, and the maximum current through the switch should not exceed 3mA. If this current is exceeded, the gate will burn out resulting in failure of the device.

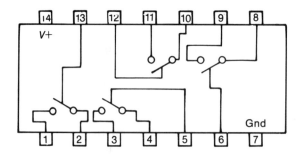

Figure 7-22. Pinout diagram for both the 4016 and 4066 quad bilateral switches

MAGNITUDE COMPARATORS

A *magnitude comparator* compares two binary numbers, and indicates whether the one number is larger than, less than, or equal to the other. **Figure 7-23** shows the logic diagram and the corresponding truth table for a one-bit magnitude comparator.

Multibit magnitude comparators are also available. For example, the 7485 is a four-bit magnitude comparator. This device will compare two four-bit binary

7
Digital Integration

numbers, and determine whether one binary number is greater than, less than, or equal to the other.

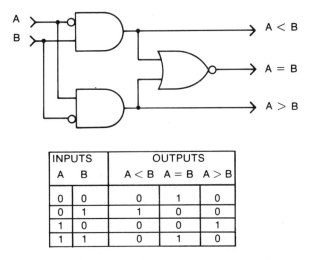

Figure 7-23. Logic diagram and corresponding truth table for a one-bit magnitude comparator

PARITY GENERATOR/CHECKERS

When used, a *parity bit* or *check bit* is automatically added to a word about to be digitally transmitted. Parity is further defined as being either even parity or odd parity.

Let's first consider a system that uses *even parity*. If the character to be sent contains an odd number of *high* bits, a *parity generator* automatically makes the parity bit *high* so that an even number of *highs* will be transmitted. If the character to be sent consists of an even number of *high* bits, the parity generator makes its bit *low*. Therefore, only an even number of *highs* again will be transmitted. In a system using even parity, the total of the *high* bits in the character and check bit will always be an even number.

Thus, in an even parity system, any character received with an odd number of *highs* is known to be erroneous. Some distortion in transmission has occurred, and an error *flags* the system. Typically, the receiver signals the transmitter to send the character again, until it's received with an even number of bits.

As you might expect, odd parity works the same way except that the parity generator makes the parity bit high only if the character to be transmitted contains an *even* number of *highs*. Likewise, the receiver accepts transmitted characters with an odd number of *highs*, and flags as a transmission error characters received with

Digital Circuits and Devices

an even number of *highs*.

The parity bit-checking system isn't foolproof. For example, suppose a transmission error causes two *high* bits to be received as two *lows* or two *low* bits to be received as two *highs*. Neither an odd nor an even parity checking system would detect such errors.

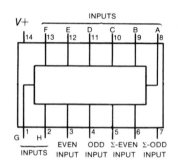

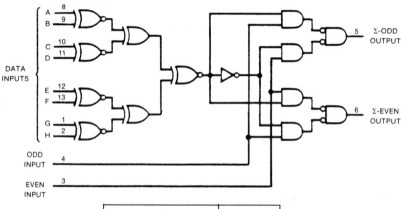

INPUTS			OUTPUTS	
Σ of H's AT A THRU H	EVEN	ODD	Σ EVEN	Σ ODD
EVEN	H	L	H	L
ODD	H	L	L	H
EVEN	L	H	L	H
ODD	L	H	H	L
X	H	H	L	L
X	L	L	H	H

Figure 7-24. Connection diagram, truth table, and logic diagram of the 74180 nine-bit parity generator/checker

7
Digital Integration

The 74180 shown in **Figure 7-24** is a nine-bit parity generator/checker. It can generate a parity bit using 8 data bits for the parity code. This device can also be used to check a 9-bit transmission from a computer or other digital device containing 8 data bits and one parity bit.

Topic Review 7-7

1. An ___ ___ ___ is a type of combination gate unit useful in designing complex gate circuits.

2. A ___ ___ gate is used to increase the noise margin of a TTL gate.

3. A ___ ___ ___ allows digital data to flow in either direction as long as the device is enabled.

4. A ___ ___ compares two binary numbers together.

5. A ___ ___ ___ can be used to check a nine-bit digital transmission consisting of eight ___ bits and one ___ bit.

Answers:

1. expandable gate function
2. Schmitt trigger
3. digital bilateral switch
4. magnitude comparator
5. nine-bit parity checker, data, parity

7-8 Summary Points

1. A half adder adds two 1-bit binary numbers together. A full adder adds three 1-bit binary numbers together. A parallel adder adds two multi-bit binary numbers together. A half subtractor subtracts two 1-bit binary numbers from each other.

2. A multiplexer is a digital device with many inputs and one output, and it can select which input will be applied to the output.

3. A demultiplexer is a digital device with one input and many outputs, and it can select which output the input data will be applied to.

Digital Circuits and Devices

4. An encoder converts a quantity into its binary equivalent.

5. A decoder interprets a binary code.

6. The three types of display sources are gas discharge tubes, LEDs, and LCDs.

7. The four basic types of display formats are: 7-segment, multiple 7-segment, dot matrix, and multiple dot matrix displays.

8. Expandable gate units allow the user to increase the number of complex gate circuits with a minimum amount of IC chips.

9. A magnitude comparator compares two binary numbers and determines if one number is larger than, smaller than, or equal to the other number.

10. A parity bit is an additional bit used by a computer to determine if an error in communication has occurred.

7-9 Chapter Progress Evaluation

1. What type of adder do you need to add 0101 to 1111?

2. Select lines are used to perform what function in a multiplexer?

3. What is the purpose of the strobe line in multiplexers and demultiplexers?

4. How many select lines are needed for an 8 to 1 multiplexer?

5. How many outputs can five select lines control in a demultiplexer?

6. In a common cathode display all cathodes are tied together and applied to what logic level?

7. A 6X9 LED dot matrix contains how many LEDs?

8. The LT pin on the 7448 is used to?

9. What will be the level of the parity bit if the parity generator is set on odd parity and the data on the inputs are 0111 0110?

Chapter 8

Latches/Flip-Flops/ Shift Registers

Objectives:

Upon completion of this chapter, you should be able to do the following:

- Describe the operation of a latch, flip-flop and shift register

- Troubleshoot any type of circuit given in this chapter with a truth table or timing diagram

- Describe the operation of a shift register used for a counter

- Develop a truth table or timing diagram for any type of circuit given in this chapter

8-1 Introduction

In this chapter you will be learning about flip-flops, latches, and shift registers. All three types of circuitry are concerned with memory; that is, they are used to store either one bit of information, as in the case of the flip-flop, or in storing multiple bits of information, such as a binary word, in the shift register. While flip-flops can be set and reset, shift registers, which are made up of combinations of flip-flops, must be reset using clock pulses.

As with a flip-flop, a latch has the ability to remain at a specific level, *high* or *low*, after having been set to that state. Flip-flops and latches will remain in their set state even after the input control pulse no longer exists. However, they will change states when the proper digital data is applied to their inputs.

8-2 RS Latches

A latch is a level triggered device that will retain the data on its outputs,

Digital Circuits and Devices

until the condition on its inputs change levels. The latch is the simplest type of single bit memory storage device. RS latches (also called the Reset/Set latch) are the basic element of storage of binary data in many digital devices. Due to its disallowed state, which causes the latch to not function logically, RS latches are not normally made in single chip form. Instead, the RS latch is combined with other RS latches and an assortment of different types of logic gates to form clock latches, data latches, and flip-flops. There are two types of RS latches. One is made from two NOR gates (RS NOR latch) and the other is made from NAND gates (RS NAND latch).

RS NOR LATCH

The RS NOR latch uses a pair of crossed-coupled NOR gates to perform its latch function. See **Figure 8-1a**. By crossed-coupled NOR gates we mean the output of NOR gate 1 acts as one input to NOR gate 2. The output of NOR gate 2 acts as one input to NOR gate 1.

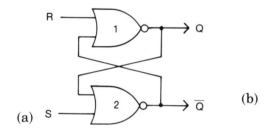

R	S	Q	$\overline{Q}$	Mode
0	0	NC	NC	hold
0	1	1	0	set
1	0	0	1	reset
1	1	0	0	disallowed

Where NC = no change in output from previous condition.

Figure 8-1 RS NOR latch with truth table and symbol

From the truth table in **Figure 8-1b**, you should note that the latch has two outputs Q and $\overline{Q}$. Therefore $\overline{Q}$ should always have the opposite condition on it, as compared to output Q. If both outputs have the same level on them, the latch is in the disallowed mode. Logic states that the levels on Q and $\overline{Q}$ must always be different; if they are the same the latch will not operate logically. Therefore, a latch should never be operated in the disallowed mode.

In order to determine the truth table for a latch, we must assume a power-up output level on Q and $\overline{Q}$. Remember that Q and $\overline{Q}$ should always have the opposite condition on them. There is no way of determining which NOR gate will switch first and assumptions must be made for each possible condition of Q and $\overline{Q}$ output.

In assumption 1, Q is *high* and $\overline{Q}$ is *low*, when power is first applied to the circuit. In assumption 2, Q is *low* and $\overline{Q}$ is *high*, when power is first applied to the circuit.

Throughout the following examples of the operations of the RS NOR latch, we

8
Latches/Flip-Flops/Shift Registers

will also assume that NOR gate 1 will switch before NOR gate 2.

Hold Mode: (R = 0 S = 0)

ASSUMPTION 1: When a *low* from $\overline{Q}$ and a *low* from input R are NORed together as shown in **Figure 8-1a**, the output from NOR gate 1 (Q) will be *high* (no change in assumed level). The *high* from output Q is now NORed to the *low* from input S and the output from NOR gate 2 ($\overline{Q}$) will be *low* (no change in assumed level). Since there were no output changes, the circuit is said to be latched.

ASSUMPTION 2: When a *high* from $\overline{Q}$ and a *low* from input R are NORed together as shown in **Figure 8-1a**, the output from NOR gate 1 (Q) will be *low* (no change in assumed level). The low from Q is now NORed to the *low* from input S and the output from NOR gate 2 ($\overline{Q}$) will be *high* (no change in assumed level). Since there were no output changes, the circuit is said to be latched.

By examination of both output assumptions, it can be concluded that when inputs R and S are *low* the output of the RS NOR latch will not change from its previous state. Therefore, the data on the output will remain the same and when this occurs the latch is said to be in the *hold mode*.

Set Mode: (R = 0, S = 1)

ASSUMPTION 1: When a *low* from $\overline{Q}$ and a *low* from input R are NORed together as in **Figure 8-1a**, the output from NOR gate 1 (Q) will be *high* (no change in assumed level). The *high* from Q is now NORed to the *high* from input S and the output from NOR gate 2 ($\overline{Q}$) will be *low* (no change in assumed level). Since there were no output changes, the circuit is said to be latched. This can be seen from the truth table in **Figure 8-1b**.

ASSUMPTION 2: When a *high* from $\overline{Q}$ and a *low* from input R are NORed together, the output from NOR gate 1 (Q) will be *low* (no change in assumed level). The *low* from Q is now NORed to the *high* from input S and the output from NOR gate 2 ($\overline{Q}$) will be *low* (change in assumed level). Since there was a change in output $\overline{Q}$, you must go through the input conditions again until the outputs do not change from their previous state. Now the *low* from $\overline{Q}$ and the *low* from input R are NORed together and the output from NOR gate 1 (Q) will be *high* (changed in previous level). The *high* from Q is now NORed to the *high* from input S and the output from NOR gate 2 ($\overline{Q}$) will be *low* (no change in previous level). Since there was a change in output Q, you must go through the input condition again. Therefore, the *low* from $\overline{Q}$ and the *low* from input R are NORed together, and the output from NOR gate 1 (Q) will be *high* (no change in previous level). The *high* from Q is now NORed to the *high* from input S, and the $\overline{Q}$ output will be *low* (no change in previous level). Since there was no change in the output states, the circuit is now latched.

From the two output assumptions made, it can be concluded that when input R is *low* and input S is *high*, Q will always be *high* and $\overline{Q}$ will always be *low*. With Q always *high* and $\overline{Q}$ *low*, it is called the *set mode* (Q is set).

Reset Mode: (R = 1, S = 0)

ASSUMPTION 1: When a *low* from $\overline{Q}$ and a *high* from input R are NORed together as in **Figure 8-1a**, the output from NOR gate 1 (Q) will be *low* (change in assumed level). The new *low* from Q is now NORed to the *low* from input S and the output from NOR gate 2 ($\overline{Q}$) will be *high* (change in assumed level). Since there was a level change on the outputs, those new output levels must be applied to the input of the NOR gates until no more changes occur in the outputs. The new *high* from $\overline{Q}$ and the *high* from input R are NORed together and the output from NOR gate 1 (Q) will be *low* (no change in previous level). The *low* from Q is NORed to the *low* from input S and the output from NOR gate 2 ($\overline{Q}$) will be *high* (no change in previous level). Since the outputs will no longer switch states, the circuit is said to be latched.

ASSUMPTION 2: When a *high* from $\overline{Q}$ and a *high* from input R are NORed together, the output from NOR gate 1 (Q) will be *low* (no change in assumed level). The *low* from (Q) is now NORed to the *low* from input S and the output from NOR gate 2 ($\overline{Q}$) will be *high* (no change in assumed level). Since there were no changes in the outputs, the circuit is said to be latched.

By comparing both examples, it can be concluded that when input R is *high* and input S is *low*, output Q will always be *low* and output $\overline{Q}$ will always be *high*. Because Q will always be *low* and $\overline{Q}$ *high*, when R = 1 and S = 0, the latch is said to be in the *reset mode*, (Q is reset).

Disallowed Mode (R = 1, S = 1)

ASSUMPTION 1: When a *low* from $\overline{Q}$ and a *high* from input R are NORed together as in **Figure 8-1a**, the output from NOR gate 1 (Q) will be *low* (change in assumed level). The new *low* from Q is now NORed to the *high* from input S and the output from NOR gate 2 ($\overline{Q}$) will be *low* (no change in assumed level). Since there was a level change, the new output conditions must be applied to the inputs of the NOR gates until no more changes occur in the outputs. When the *low* from $\overline{Q}$ and the *high* from input R are NORed together, the output from NOR gate 1 (Q) will be *low* (no change in previous level). The *low* from Q is NORed to the *high* from input S and the

8
Latches/Flip-Flops/Shift Registers

output from NOR gate 2 ($\overline{Q}$) will be *low* (no change in previous output). Since the outputs will no longer switch states, the circuit is said to be latched.

ASSUMPTION 2: When a *high* from $\overline{Q}$ and a *high* from input R are NORed together, the output from NOR gate 1 (Q) wil be *low* (no change in assumed level). The low from Q is now NORed to the *high* from input S and the output from NOR gate 2 ($\overline{Q}$) will be *low* (change in assumed level). Since there was a change in the output level, the new levels must be applied to the inputs of the NOR gates until no more changes occur in the outputs. The new *low* from $\overline{Q}$ and the *high* from input R are NORed together and the output from NOR gate 1 (Q) will be *low* (no change in previous level). The *low* from (Q) is now NORed to the *high* from input S and the output from NOR gate 2 ($\overline{Q}$) will be *low* (no change in previous level). Since there were no changes in the outputs, the circuit is said to be latched.

By comparing both examples it can be concluded that when input R is *high* and input S is *high*, the output Q will always be *low* and output $\overline{Q}$ will always be *low*. This condition is not a logical operation because logic states that Q and $\overline{Q}$ must always have opposite conditions. If the outputs of the latch are the same, circuits connected to the outputs of the latch will not work properly, making the latch useless to the circuit. When both outputs are at the same level, it is known as the *disallowed mode*; this mode should not be used.

RS NAND LATCH

The RS NAND latch shown in **Figure 8-2a**, uses a pair of crossed-coupled NAND gates to perform its latch function. As you might have guessed, since the NAND gate operates differently than the NOR gate, all conditions for each input will produce the opposite results. Since this condition occurs, inputs R and S have negation signs over them to indicate reverse operation of the RS NAND latch. Also **Figure 8-2b** shows the symbol for the RS NAND latch while **Figure 8-2c** expresses its truth table.

In assumption 1, Q is *high* and $\overline{Q}$ is *low* when power is first applied to the circuit. In assumption 2, Q is *low* and $\overline{Q}$ is *high* when power is first applied to the circuit. Another assumption we will make is that NAND gate 1 will always switch first and NAND gate 2 will switch second in the following examples. **Figure 8-2a** and **8-2c** can be referred to when discussing the different modes of operation of this latch.

Disallowed Mode ($\overline{R} = 0, \overline{S} = 0$)

ASSUMPTION 1: When a *low* from $\overline{Q}$ and a *low* from input $\overline{S}$ are NANDed together, the output from NAND gate 1 (Q) will be *high* (no change in assumed level).

Digital Circuits and Devices

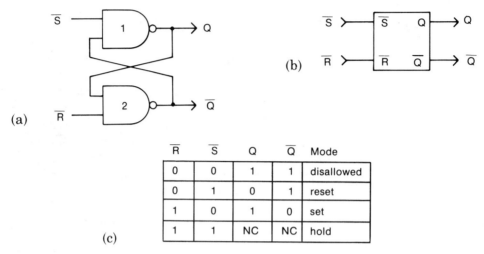

Figure 8-2 RS NAND latch with a) internal diagram, b) symbol, and c) truth table

The *high* from output Q is now NANDed to the *low* from input $\overline{R}$ and the output from NAND gate 2 ($\overline{Q}$) will be *high* (change in assumed level). Since there was a change in the output level, the new levels must be applied to the inputs, until the outputs of the circuit no longer change. The new *high* from $\overline{Q}$ and the *low* from input $\overline{S}$ are NANDed together, and the output from NAND gate 1 (Q) will be *high* (no change in previous level). The *high* from Q is now NANDed to the *low* from input $\overline{R}$ and the output from NAND gate 2 ($\overline{Q}$) will be *high* (no change in previous level). Since there were no level changes, the circuit is said to be latched.

ASSUMPTION 2: When a *high* from $\overline{Q}$ and a *low* from input $\overline{S}$ are NANDed together, the output from NAND gate 1 (Q) will be *high* (change in assumed level). The *high* from Q is now NANDed to the *low* from input $\overline{R}$ and the output from NAND gate 2 ($\overline{Q}$) will be *high* (no change in assumed level). Since there was a change in the output level, the levels must be applied to the inputs of the NAND gates until no more level changes occur. The new *high* from $\overline{Q}$ and the *low* from $\overline{S}$ are NANDed together and the output from NAND gate 1 (Q) will be *high* (no change in previous level). The *high* from Q is now NANDed to the *low* from input $\overline{R}$ and the output from NAND gate 2 ($\overline{Q}$) will be *high* (no change in previous level). Since there was no change in levels, the circuit is said to be latched.

By comparing both examples, it can be concluded that when inputs $\overline{R}$ and $\overline{S}$ are *low* the outputs from the RS NAND latch will both be *high*. Since Q and $\overline{Q}$ must have different levels on them to operate properly, the latch is in the *disallowed mode*.

Latches/Flip-Flops/Shift Registers

Reset Mode ($\overline{R} = 0, \overline{S} = 1$)

ASSUMPTION 1: When a *low* from $\overline{Q}$ and a *high* from input $\overline{S}$ are NANDed together, the output from NAND gate 1 (Q) will be *high* (no change in assumed level). The *high* from Q is now NANDed to the *low* from input $\overline{R}$ and the output from NAND gate 2 ($\overline{Q}$) will be *high* (change in assumed level). Since there was a change in levels, the new levels must be applied to the inputs until the outputs of the circuit no longer change. The new *high* from $\overline{Q}$ and the *high* from input $\overline{S}$ are NANDed together and the output from NAND gate 1 (Q) will be *low* (change in previous level). The *low* from Q is now NANDed to the *low* from input $\overline{R}$ and the output from NAND gate 2 ($\overline{Q}$) will be *high* (no change in previous level). Since the Q output changed levels, the output levels must be applied again. The *high* from $\overline{Q}$ is now NANDed to the *high* from input $\overline{S}$, and the Q output will be *low* (no change from previous level). The *low* from Q and the *low* from $\overline{R}$ are now NANDed yielding a *high* $\overline{Q}$ output (no change in previous level). Since there will be no more level changes, the circuit is said to be latched.

ASSUMPTION 2: When a *high* from $\overline{Q}$ and a *high* from input $\overline{S}$ are NANDed together, the output from NAND gate 1 (Q) will be *low* (no change in assumed level). Because the *low* from Q is now NANDed to the *low* from input $\overline{R}$, the output from NAND gate 2 ($\overline{Q}$) will be *high* (no change in assumed level). Since there were no changes in levels, the circuit is said to be latched.

By comparing both examples, it can be concluded that when input $\overline{R}$ is *low* and $\overline{S}$ is *high*, Q will be *low* and $\overline{Q}$ will be *high*. When Q is *low* and $\overline{Q}$ is *high*, the RS NAND latch is in the *reset mode*.

Set Mode ($\overline{R} = 1, \overline{S} = 0$)

ASSUMPTION 1: When a *low* from $\overline{Q}$ and a *low* from input $\overline{S}$ are NANDed together, the output from NAND gate 1 (Q) will be *high* (no change in assumed level). The *high* from Q is now NANDed to the *high* from input $\overline{R}$ and the output from NAND gate 2 ($\overline{Q}$) will be *low* (no change in assumed level). Since there were no changes in levels, the circuit is said to be latched.

ASSUMPTION 2: When the *high* from $\overline{Q}$ and the *low* from input $\overline{S}$ are NANDed together, the output from NAND gate 1 (Q) will be *high* (change in assumed level). The new *high* from Q is now NANDed to the *high* from input $\overline{R}$, the output from NAND gate 2 ($\overline{Q}$) will be *low* (change in assumed level). Since there was a

Digital Circuits and Devices

change in levels, the levels must be applied to the NAND inputs until the output levels no longer change. The new *low* from $\overline{Q}$ is NANDed to the *low* from input $\overline{S}$ and the output from NAND gate 1 (Q) will be *high* (no change in previous level). The *high* from Q is NANDed with the *high* from input $\overline{R}$ and the output from NAND gate 2 ($\overline{Q}$) will be *low* (no change in previous level). Since there was no change in the levels, the circuit is said to be latched.

By comparing both examples it can be concluded that when input $\overline{R}$ is *high* and $\overline{S}$ is *low*, output Q will be *high* and $\overline{Q}$ will be *low*. When Q is *high* and $\overline{Q}$ is *low* the RS NAND latch is in the *set mode*.

Hold Mode ($\overline{R} = 1, \overline{S} = 1$)

ASSUMPTION 1: When the *low* from $\overline{Q}$ and the *high* from input $\overline{S}$ are NANDed together, the output from NAND gate 1 (Q) will be *high* (no change in assumed level). The *high* from Q is now NANDed to the *high* from input $\overline{R}$, the output from NAND gate 2 ($\overline{Q}$) will be *low* (no change in assumed level). Since there was no change in levels, the circuit is said to be latched.

ASSUMPTION 2: When the *high* from $\overline{Q}$ and the *high* from input $\overline{S}$ are NANDed together, the output from NAND gate 1 (Q) will be *low* (no change in assumed level). The *low* from Q is now NANDed to the *high* from input $\overline{R}$ and the output from NAND gate 2 ($\overline{Q}$) will be *high* (no change in assumed level). Since there was no change in levels, the circuit is said to be latched.

By comparing both examples, it can be concluded that when $\overline{R}$ and $\overline{S}$ are *high* the outputs of the circuit will not change. When the input condition does not allow the output levels to change, the RS NAND latch is in the *hold mode*.

Before proceeding to the RS clocked latches, be sure you know the operations of both the RS NOR latch and RS NAND latch. These latches are the basic building blocks of all other latches and flip-flops.

Topic Review 8-2

1. When $\overline{R} = 0$ and $\overline{S} = 1$ in the RS NAND latch, Q = _____ and $\overline{Q}$ = _____.
2. The RS NOR latch will be in the _____ mode when R = 1 and S = 1.

Answers:

1. 0, 1
2. disallowed

8-3 Clocked RS Latches

A clocked RS latch is a device that retains the data on its outputs, until the proper clock level and input conditions cause the outputs to change their levels.

CLOCKED RS NOR LATCH

The clocked RS NOR latch is made up from a RS NOR latch and two clocked AND gates. Before proceeding with the operation of the clocked RS NOR latch, see **Figure 8-3**.

(a)

(b)

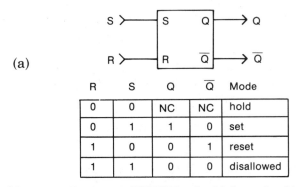

Figure 8-3 RS NOR latch with its truth table

In the operation of the clocked RS NOR latch, we will be using the truth table for the RS NOR latch. **Figure 8-3a** shows the symbol for the RS NOR latch. **Figure 8-3b** shows the truth table for the same latch. **Figure 8-4a** shows the block diagram for the clocked RS NOR latch. Notice the addition of two AND gates at the input to this RS NOR latch. **Figure 8-4b** shows the symbol for the clocked RS NOR latch and **Figure 8-4c** shows its truth table. Notice also the addition of a column labeled "CK" for clock and also the addition of short pulses within that column.

Disable Mode (CK = 0, R = X, S = X)

Referring to **Figure 8-4a**, when the clock input is *low*, the outputs of both AND gates will be *low*, no matter what levels are on the other inputs. Since both outputs from the AND gates are *low*, a *low* is applied to the R and S inputs of the internal RS NOR latch. When a *low* is placed on both the R and S inputs of the RS NOR latch, the outputs of the latch will not change from their previous levels. Since the inputs of the AND gates have no affect on the output of the circuit, the latch is said to be in the

Digital Circuits and Devices

disable mode (inputs do not affect the output).

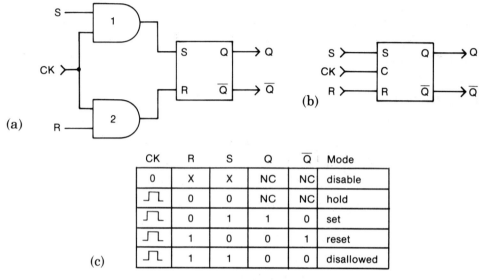

Where X = level does not matter.

Where ⎍ = active *high* clock which enables the inputs.

Figure 8-4 Clocked RS NOR latch with symbol and truth table

Hold Mode (CK = ⎍, R = 0, S = 0)

When the clock level is *high* and input S is *low*, the output of AND gate 1 will be *low*. At the same time input R is *low*, the output of AND gate 2 will be *low*. The output from AND gate 1 is applied to the S input and the output from AND gate 2 is applied to the R input of the internal RS NOR latch. When a *low* is on input S and a *low* is on input R of the RS NOR latch, outputs Q and $\overline{Q}$ will not change from their previous levels. The latch is said to be in its *hold mode* (no change in the output when the clock is enabled).

Set Mode (CK = ⎍, R = 0, S = 1)

When the clock level is *high* and input S is *high*, the output of AND gate 1 will be *high*. At the same time input R is *low*, the output from AND gate 2 will be *low*. The output from AND gate 1 is applied to the S input and the output from AND gate 2 is applied to the R input of the internal RS NOR latch. When a *high* is on input S and a *low* is on input R of the RS NOR latch, output Q will go *high* and output $\overline{Q}$ will go

8 Latches/Flip-Flops/Shift Registers

low. The latch is said to be in its *set mode* (Q is set).

Reset Mode (CK = ⎍, R = 1, S = 0)

When the clock level is *high* and input S is *low*, the output of AND gate 1 will be *low*. At the same time input R is *high*, the output from AND gate 2 will be *high*. The output from AND gate 1 is applied to the S input and the output from AND gate 2 is applied to the R input of the internal RS NOR latch. When a *low* is on input S and a *high* is on input R of the RS NOR latch, output Q will go *low* and output $\overline{Q}$ will go *high*. The latch is said to be in its *reset mode* (Q is reset).

Disallowed Mode (CK = ⎍, R = 1, S = 1)

When the clock level is *high* and input S is *high*, the output of AND gate 1 will be *high*. At the same time input R is *high*, the output of AND gate 2 will be *high*. The output from AND gate 1 is applied to the S input and the output from AND gate 2 is applied to the R input of the internal RS NOR latch. When a *high* is on input S and a *high* is on input R of the RS NOR latch, output Q will go *low* and output $\overline{Q}$ will go *low*. The latch is said to be in its *disallowed mode* (a non-logical operation).

From the truth table and the explanations of the operation of the clocked RS NOR latch it can be concluded that the clock must be *high* in order to allow the latch to change its outputs. Whenever the clock is *low*, the ability for the latch to change its outputs will not exist.

CLOCKED RS NAND LATCH

The clocked RS NAND latch is made up of an RS NAND latch and two clocked NAND gates. Before proceeding with the operation of the clocked RS NAND latch, see **Figure 8-5**.

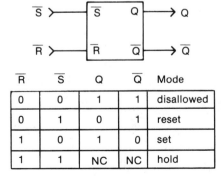

Figure 8-5 RS NAND latch with truth table

Digital Circuits and Devices

In the operation of the clocked RS NAND latch, we will be using the truth table for the RS NAND latch. **Figure 8-5a** shows the symbol for the RS NAND latch. **Figure 8-5b** shows the truth table for the same latch. **Figure 8-6a** shows the block diagram for the clocked RS NAND latch. Notice the addition of two NAND gates at the input to this RS NAND latch. **Figure 8-6b** shows the symbol for the clocked RS NAND latch and **Figure 8-6c** shows its truth table. Notice also the addition of a column labeled "CK" for clock and also the addition of short pulses within that column. We will now proceed with the discussion of the RS NAND latch while referring to **Figure 8-6a and 8-6b**.

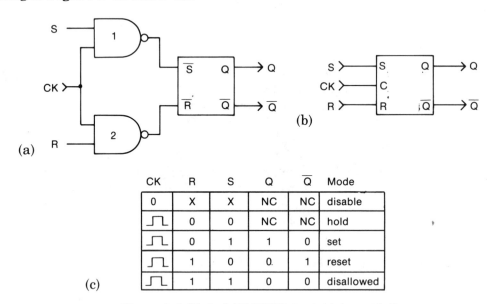

Figure 8-6 Clocked RS NAND latch a) internal diagram, b) symbol, and c) truth table

Disable Mode (CK = 0, R = X, S = X)

When the clock input is *low*, the output of both NAND gates will be *high*, no matter what levels are on the other inputs. Since both outputs from the NAND gates are *high*, a *high* is applied to the $\overline{R}$ and $\overline{S}$ inputs of the internal RS NAND latch. When a *high* is placed on both $\overline{R}$ and $\overline{S}$ inputs of the RS NAND latch, the outputs of the latch will not change from their previous levels. Since the inputs of the NAND gates have no effect on the output of the circuit, the latch is said to be in the *disable mode* (inputs do not affect the output without a clock).

Hold Mode (CK = ⊓, R = 0, S = 0)

When the clock level is *high* and input S is *low*, the output from NAND gate 1 will be *high*. At the same time input R is *low*, the output of NAND gate 2 will be *high*. The output from NAND gate 1 is applied to the $\overline{S}$ input and the output from NAND gate 2 is applied to the $\overline{R}$ input of the internal RS NAND latch. When a *high* is on input $\overline{S}$ and a *high* is on input $\overline{R}$ of the RS NAND latch, outputs Q and $\overline{Q}$ will not change from their previous levels. The latch is said to be in its *hold mode* (no change in the outputs with the proper clock).

Set Mode (CK = ⎍, R = 0, S = 1)

When the clock level is *high* and input S is *high* the output of NAND gate 1 will be *low*. At the same time input R is *low*, the output from NAND gate 2 will be *high*. The output from NAND gate 1 is applied to the $\overline{S}$ input and the output from NAND gate 2 is applied to the $\overline{R}$ input of the internal RS NAND latch. When a *low* is on input $\overline{S}$ and a *high* is on input $\overline{R}$ of the RS NAND latch, output Q will go *high* and output $\overline{Q}$ will go *low*. The latch is said to be in its *set mode* (Q is set).

Reset Mode (CK = ⎍, R = 1, S = 0)

When the clock level is *high* and input S is *low*, the output of NAND gate 1 will be *high*. At the same time input R is *high*, the output from NAND gate 2 will be *low*. The output from NAND gate 1 is applied to the $\overline{S}$ input, and the output from NAND gate 2 is applied to the $\overline{R}$ input of the internal RS NAND latch. When a *high* is on input $\overline{S}$ and a *low* is on input $\overline{R}$ of the RS NAND latch, output Q will go *low* and output $\overline{Q}$ will go *high*. The latch is said to be in its *reset mode* (Q is reset).

Disallowed Mode (CK = ⎍, R = 1, S = 1)

When the clock level is *high* and input S is *high*, the output of NAND gate 1 will be *low*. At the same time input R is *high*, the output of NAND gate 2 will be *low*. The output from NAND gate 1 is applied to the $\overline{S}$ input and the output from NAND gate 2 is applied to the $\overline{R}$ input of the internal RS NAND latch. When a *low* is on input $\overline{S}$ and a *low* is on input $\overline{R}$ of the RS NAND latch, output Q will go *high* and output $\overline{Q}$ will go *high*. The latch is said to be in its *disallowed mode* (a non-logical operation).

From the truth table and the explanations of the operation of the clocked RS NAND latch, it can be concluded that the clock must be *high* in order to allow the latch function to work properly. Whenever the clock is *low*, the ability for the latch to change its levels will not exist. It should also be noted that the truth tables for the clocked RS NOR latch and the clocked RS NAND latch, are the same. The reason for this is that the inputs of the RS NAND latch are controlled by the NAND gates. The NAND gates will invert each input level and cause the internal NAND latch to operate like an RS NOR latch.

Digital Circuits and Devices

Topic Review 8-3

1. The clocked RS NAND latch will be in the _____ mode when R = 0 and S = 0.

2. The clocked RS NAND latch will be in the _____ mode when R = 1 and S = 0.

3. The circuit difference that makes the clocked RS NAND latch operation the same as the clocked RS NOR latch is the _____ _____ gates.

Answers:

1. hold
2. Reset
3. input NAND

8-4 Data Latches

A data latch is a device that will retain the information on its output until the enable level and input level causes the data to change.

The data latch is nothing more than a clocked RS latch that has its R input inverted as compared to its S input. The input for the data latch is called D (Data) and there is only one output, Q.

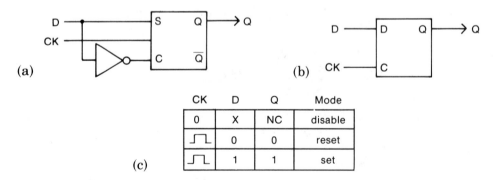

Figure 8-7 Data latch a) internal diagram, b) symbol, and c) truth table

The internal diagram of the data latch is shown in **Figure 8-7a**. Its symbol is shown in **Figure 8-7b** and the truth table for the data latch is expressed in **Figure 8-7c**. **Figure 8-7a** and **8-7c** should be referred to while reading about the various operational modes of this particular latch.

8 Latches/Flip-Flops/Shift Registers

DISABLE MODE (CK = 0, D = X)

When CK is *low*, the clock input for the internal RS latch is *low*. This will disable the internal RS latch no matter what is on its inputs. The latch is then in its *disable mode* (output level will not change no matter what is on its input).

RESET MODE (CK = ⎍ , D = 0)

When CK is *high*, the clock input for the internal RS latch is *high*. This will enable the RS latch so that it can change its outputs. At the same time a *low* from input D is applied to the S input and a *high* is applied to input R. When S is *low* and R is *high* the output of the internal RS latch will produce a *low* on Q and a *high* on $\overline{Q}$. The latch is then in the *reset mode* (Q is reset).

SET MODE (CK = ⎍ , D = 1)

When CK is *high*, the clock input for the internal RS latch is *high*. This will enable the RS latch so that it can change its outputs. At the same time a *high* from input D is applied to the S input and a *low* is applied to input R. When S is *high* and R is *low* the output of the internal RS latch will produce a *high* on Q and a *low* on $\overline{Q}$. The latch is then in the *set mode* (Q is set).

From the truth table and the explanation of the circuit it can be concluded that the data latch will display the data on its output and will retain it until the enable input goes *high*, and the data input causes data to change. Data will not change when the enable input is *low*. It should also be noted that there is no disallowed mode because input S and input R are always in the opposite condition when applied to the internal clock RS latch.

7477 QUAD DATA LATCH

The 7477, shown in **Figure 8-8a** with the truth table in **Figure 8-8b**, is a quad data latch containing four data latches. Note the enable lines for two latches are connected together which decreases the number of pins on the IC chip. The operation of the quad data latch is the same as the single data latch, except two data latches are enabled at the same time.

Data latches are used as simple memory devices to store binary data in small digital circuits until the data can be used in the circuit. Because of the latch itself, normally only eight can be placed on an IC chip, so the amount of data you can store with one IC latch is eight bits. If more data needs to be retained, memory chips are normally used.

Digital Circuits and Devices

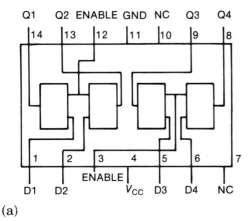

INPUTS		OUTPUTS	
D	G	Q	$\overline{Q}$
L	H	L	H
X	L	Q_0	$\overline{Q_0}$

H = High Level (steady state)
L = Low Level (steady state)
X = Don't Care
↑ = Transition from low to high level
Q_0 = The level of Q before the indicated steady state input conditions were established

(a) (b)

(Courtesy National Semiconductor Corp.)

Topic Review 8-4

1. A data latch allows data to change whenever the enable is _____.

2. No change from the previous level will be the condition of a data latch, when the enable input is _____.

Answers:

1. *high*
2. *low*

8-5 RS Master Slave Flip-Flops

An RS master slave flip-flop is a digital device that will retain its output data until the proper clock edge and input conditions cause the output to change. The RS master slave flip-flop contains two clocked RS latches and one inverter.

Before describing the operation of the RS master slave flip-flop, note the inverter between the clock inputs of the master and slave sections in **Figure 8-9a**. When a *high* clock level is applied to the clock input, the master section is enabled and data can change on the outputs of the master section. At this time the slave section is disabled and any previous data that was on the outputs will remain. During the transition from a *high* to a *low* of the clock, the master section will become disabled so that data on the master section outputs will remain unchanged. Also during the transition, the inverter will place a *high* on the clock input of the slave section. The

206

8
Latches/Flip-Flops/Shift Registers

data on the master section will be applied to the inputs of the slave section. Depending upon the input data, the output level in the slave section will change. Since the master section is now disabled, any further input changes on the flip-flop will not be seen until the clock level goes *high*. The symbol for this particular flip-flop is shown in **Figure 8-9b** and its truth table is expressed in **figure 8-9c**. **Figure 8-9a** will be used in discussing the various modes of operation of this particular flip-flop.

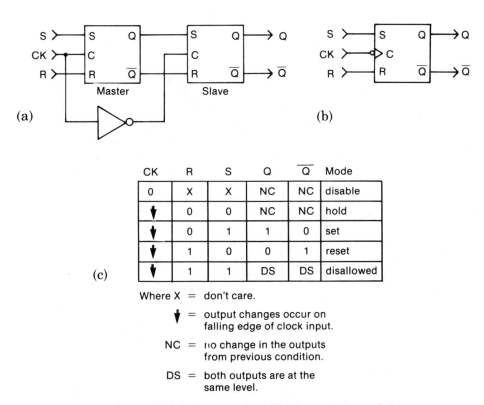

Figure 8-9 RS master slave flip-flop with (b) symbol and (c) truth table

DISABLE MODE (CK = 0, R = X, S = X)

When the clock is *low*, the master latch is disabled and the data previously on the Q and $\overline{Q}$ outputs will not change. At the same time, the output from the master

Digital Circuits and Devices

section of the flip-flop is applied to the slave section. Since at this time the slave section is enabled, the data on the output of the slave section will be the same as the data on the master section. Even if the R and S inputs to the flip-flop change, the output of the flip-flop will not change. When the inputs do not affect the outputs and the clock is *low*, the flip-flop is said to be in its *disable mode* (inputs have no control over the outputs).

HOLD MODE (CK = ↧ , R = 0, S = 0)

When the clock is *high*, R is *low* and S is *low*, the output of the master latch will not change from its previous levels. At this time the slave latch is disabled, so no change will occur in its output levels. During the clock transition from a *high* to a *low*, the master latch becomes disabled and its outputs will no longer change. The transition now enables the clock to the slave latch. Whatever levels were on the outputs of the master latch are now applied to the inputs of the slave latch. The inputs on the slave latch will produce the same outputs as those that were on the outputs of the master latch. Since the master latch did not change its data, the output of the slave latch will not change. Also, since the master latch is disabled at this time, data cannot change until the clock goes *high* again. When the output levels of a flip-flop do not change from their previous states, with the proper clock pulse, the flip-flop is said to be in the *hold mode* (output does not change with proper clock pulse).

SET MODE (CK = ↧ , R = 0, S = 1)

When the clock is *high*, input R is *low*, and S is *high*, the master latch will produce a *high* on Q and a *low* on $\overline{Q}$. At this time the slave latch is disabled. When the transition of the clock pulse goes from a *high* to a *low*, the master section will become disabled. The output data on the master latch will not change until the next *high* clock. The *high* from Q is applied to input S and the *low* from $\overline{Q}$ is applied to the R input of the slave latch. At this time the clock is *high* on the slave latch which will allow the R and S inputs to cause output Q to go *high* and $\overline{Q}$ to go *low*. The output level will not change again until another clock pulse is received. When Q is *high* and $\overline{Q}$ is low the flip-flop is said to be in the *set mode* (Q is set).

RESET MODE (CK = ↧ , R = 1, S = 0)

When the clock is *high*, input R is *high*, and S is *low*, the master latch will produce a *low* on Q and a *high* on $\overline{Q}$. At this time, the slave latch is disabled. When the transition of the clock pulse goes from a *high* to a *low*, the master section becomes disabled. The *low* from Q is applied to input S and the *high* from $\overline{Q}$ is applied to the R

Latches/Flip-Flops/Shift Registers

input of the slave latch. At this time the clock is *high*, and its inputs R and S will cause output Q to go *low* and $\overline{Q}$ to go *high*. The output level now will not change again until another clock pulse is received. When Q is *low* and $\overline{Q}$ is *high*, the flip-flop is said to be in the *reset mode* (Q is reset).

DISALLOWED MODE (CK = ↓ , R = 1, S = 1)

When the clock is *high*, input R is *high*, and input S is high, the master latch will produce a *high* on both outputs (normally, flip-flops are made with NAND latches because only one type of gate is required). At this time, the slave latch is disabled. When the transition of the clock occurs, the master latch becomes disabled. The *high* from Q and $\overline{Q}$ are then applied to inputs S and R. Now the clock input to the slave latch is enabled and along with the input levels on R and S, the output levels of the slave latch will both be *high*. When the level on both outputs of a flip-flop are the same, the flip-flop is said to be in the *disallowed mode* (a non-logical output function).

From the truth table and the explanations of the flip-flop operations, you should note that for every clock pulse only one output change can occur. Note also that the disallowed mode still remains. Because of this state, RS master slave flip-flops are not usually made in single chip form. The JK master slave flip-flop changes the disallowed mode into a toggle mode and is used in most any type of digital circuit that requires a flip-flop.

In the symbol for the RS flip-flop, the notch inside the block represents that the device is edge triggered for a flip-flop. The circle on the outside of the block represents that the output triggers on the *high* to *low* transition.

Topic Review   8-5

1. When the R and S inputs to an RS master slave flip-flop are *high* and there is no clock pulse, the flip-flop is said to be in the _____ mode.

2. When the outputs of an RS flip-flop are the same, it is said to be in the _____ mode.

3. When the clock level on an RS flip-flop is *high*, the internal _____ latch is enabled.

Answers:

1. disable
2. disallowed
3. master

Digital Circuits and Devices

8-6 JK Master Slave Flip-Flop

The JK master slave flip-flop is one of the most important digital devices that has ever been developed. **The JK flip-flop is an essential part of every type of digital counter circuits today.** A JK flip-flop contains two clocked RS latches, two AND gates and one inverter.

Basically the JK master slave flip-flop works the same as the RS master slave flip-flop except that the disallowed mode has been modified to make it a toggle mode.

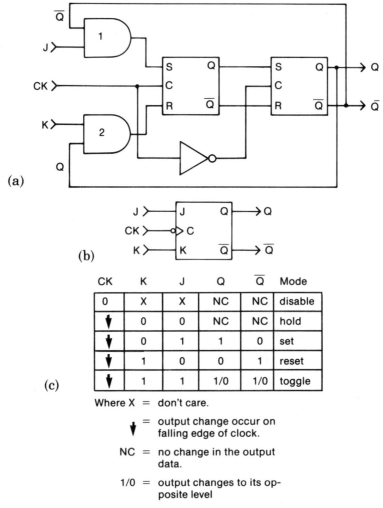

CK	K	J	Q	$\overline{Q}$	Mode
0	X	X	NC	NC	disable
↓	0	0	NC	NC	hold
↓	0	1	1	0	set
↓	1	0	0	1	reset
↓	1	1	1/0	1/0	toggle

Where X = don't care.

↓ = output change occur on falling edge of clock.

NC = no change in the output data.

1/0 = output changes to its opposite level

Figure 8-10 JK master slave flip-flop a) internal diagram, b) symbol and c) truth table

8
Latches/Flip-Flops/Shift Registers

The toggle mode in the JK flip-flop is produced by ANDing the $\overline{Q}$ output with the J input. Also, the Q output is ANDed to the K input. Even if the J and K inputs are both *high*, when ANDed to the Q and $\overline{Q}$ outputs, the inputs to the master latch will never be the same. As in the description of latches, the output from Q and $\overline{Q}$ must be assumed in order to determine how the inputs will affect the output.

Figure 8-10a shows the internal diagram of the JK master slave flip-flop. It will be used in the discussion of the various modes of operation for this particular flip-flop. **Figure 8-10b** shows the symbol used to represent the JK flip-flop and **Figure 8-10c** is the expression of its truth table.

DISABLE MODE (CK = 0, J = X, K = X)
Assumed outputs (Q = 0, $\overline{Q}$ = 1)

When the clock is *low*, the master latch in the flip-flop is disabled and no matter what is on the inputs, the outputs of the master latch will not change. The clock of the slave latch is enabled at this time, but because the outputs of the master latch did not change, the outputs of the slave latch will not change. When the inputs have no affect on the outputs, without the proper clock pulse, the flip-flop is said to be in the *disable mode*.

HOLD MODE (CK = ↓, J = 0, K = 0)
Assumed outputs (Q = 0, $\overline{Q}$ = 1)

When the clock is *high*, input J is *low*, and input K is *low*, the output of AND gate 1 will be *low* and the output of AND gate 2 will be *low*. The output of AND gate 1 will be applied to input S of the master latch. The output of AND gate 2 will be applied to input R of the master latch. With the input levels given, the Q output will be *low* and $\overline{Q}$ will be *high* (no change from previous levels). The clock input of the slave latch will be disabled at this time. When the clock pulse goes *low*, the master latch will become disabled. The clock now enables the slave latch and the S input will be *low* and the R input will be *high*. The output of Q will be *low* and $\overline{Q}$ will be *high* on the slave latch (no change from previous output levels). When the proper clock pulse is received with J and K inputs both *low*, the output levels of the flip-flop will not change, as said to be in the *hold mode* (no change in outputs with proper clock pulse).

SET MODE (CK = ↓, J = 1, K = 0)
Assumed outputs (Q = 0, $\overline{Q}$ = 1)

When the clock level is *high*, J input is *high*, and K input is *low*, the output of AND gate 1 will be *high* and AND gate 2 will be *low*. The output of AND gate 1 will

Digital Circuits and Devices

be applied to input S of the master latch. The output of AND gate 2 will be applied to input R of the master latch. With the input level given, the Q output will be *high* and $\overline{Q}$ will be *low*. The clock input of the slave latch will be disabled at this time. When the clock pulse goes *low*, the master latch will become disabled. The clock now enables the slave latch and the S input will be *high* and the R input will be *low*. The output of Q will be *high* and $\overline{Q}$ will be *low*, in the slave latch. When J is *high* and K is *low* and the proper clock pulse is received, the Q output will be *high* and $\overline{Q}$ will be *low*. When Q is *high* and $\overline{Q}$ is *low* on the JK flip-flop and the proper clock pulse is received, the JK is in the *set mode* (Q is set).

RESET MODE (CK = ↓ , J = 0, K = 1)
Assumed outputs (Q = 1, $\overline{Q}$ = 0)

When the clock is *high*, input J is *low*, and input K is *high*, the output of AND gate 1 will be *low* and AND gate 2 will be *high*. The output of AND gate 1 will be applied to input S of the master latch. The output of AND gate 2 will be applied to input R of the master latch. With the input levels given, the Q output will be *low* and $\overline{Q}$ will be *high*, in the master latch. The clock input of the slave latch will be disabled at this time. When the clock pulse goes *low*, the master latch will become disabled and the S will be *low* while the R input will be *high*, in the slave latch. The output of Q will be *low* and $\overline{Q}$ will be *high* on the slave latch. When J is *low*, K is *high*, and the proper clock pulse is received, the Q output will be *low* and $\overline{Q}$ will be *high*. When Q is *low* and $\overline{Q}$ is *high*, the flip-flop is said to be in the *reset mode* (Q is reset).

TOGGLE MODE (CK = ↓ , J = 1, K = 1)
Assumed outputs (Q = 0, $\overline{Q}$ = 1)

When the clock is *high*, input J is *high*, and input K is *high*, the output of AND gate 1 will be *high* and the output of AND gate 2 will be *low*. The inputs just given will cause the master latch to produce a *high* on Q and a *low* on $\overline{Q}$. At this time, the slave latch is disabled so there will be no change in the output. When the clock pulse goes *low*, the master latch becomes disabled. The Q output is applied to the S input and the $\overline{Q}$ of the master latch is applied to the R input of the slave latch. The Q output will be *high* and the $\overline{Q}$ output of the slave latch will be *low*. So when J and K are both *high* and the previous output of Q was *low* and $\overline{Q}$ was *high*, the outputs of the JK will switch states.

Now the assumed outputs are Q is *high* and $\overline{Q}$ is *low*. When the clock is *high*, input J is *high* and input K is *high*. The output of AND gate 1 will be *low* and the output of AND gate 2 will be *high*. The inputs just given will cause the master latch to produce a *high* on Q and a *low* on $\overline{Q}$. At this time the slave latch is disabled, so

8
Latches/Flip-Flops/Shift Registers

there will be no change in the output levels. When the clock pulse goes *low*, the master latch becomes disabled. The Q output is applied to the S input and the $\overline{Q}$ of the master latch is applied to the R input of the slave latch. The Q output will be *low* and the $\overline{Q}$ output of the slave latch will be *high*. So when J and K are both *high*, and the previous output of Q was *high* and $\overline{Q}$ was low in the JK flip-flop, the outputs of the JK will switch levels. It is very important to note that when J and K are *high*, in the JK flip-flop and a clock pulse is received, the outputs of the flip-flop will toggle to their opposite levels. Since the outputs toggle each time the clock pulse is received, the flip-flop is in the *toggle mode* (outputs toggle). This feature has made it possible for counters to be made very simple in digital circuitry.

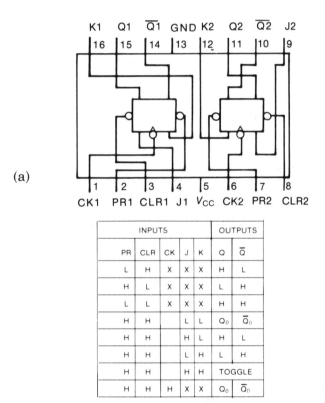

(a)

(b)

INPUTS					OUTPUTS	
PR	CLR	CK	J	K	Q	$\overline{Q}$
L	H	X	X	X	H	L
H	L	X	X	X	L	H
L	L	X	X	X	H	H
H	H		L	L	Q_0	$\overline{Q}_0$
H	H		H	L	H	L
H	H		L	H	L	H
H	H		H	H	TOGGLE	
H	H	H	X	X	Q_0	$\overline{Q}_0$

Figure 8-11 7476 dual JK flip-flop a) internal diagram and b) truth table
(Courtesy National Semiconductor Corp.)

Figure 8-11a shows the 7476 dual JK flip-flop internal diagram with preset and clear. **Figure 8-11b** shows its truth table. The chip contains two separate JK flip-flops. Each flip-flop has an active *low* preset pin, which when low will cause the Q

213

Digital Circuits and Devices

output to set and $\overline{Q}$ to reset. Each flip-flop also contains an active *low* clear pin, which when *low* will cause the Q output to reset and $\overline{Q}$ to set. Note for normal operation that the preset and clear pins should be tied *high*. One last condition to note is that the preset and clear pins will override any input conditions in the flip-flop and do not require a clock pulse.

Topic Review 8-6

1. The circuitry differences between the JK and RS flip-flops are: J and $\overline{Q}$ are _____, and K and Q are _____.

2. The disallowed mode of the RS flip-flop has been modified so that in the JK flip-flop it is called a _____ mode.

Answers:

1. ANDed, ANDed
2. toggle

8-7 Data Flip-Flops

The data flip-flop retains its output data until the input data and clock pulse cause the output to change. The operation of the data flip-flop is the same as the data latch except that data changes on the *edge* of the enable line instead of during the enable level.

One such type of flip-flop is the 74174 hex D flip-flop with clears as shown in **Figure 8-12.** This IC contains six data flip-flops, with a common clock input and common clear input. **Figure 8-12a** shows its connection diagram, **8-12b** shows its logic diagram, and **8-12c** its truth table.

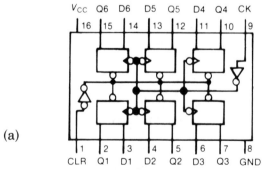

(a)

Figure 8-12 74174 hex D flip-flop a) connection diagram, b) logic diagram, and c) truth table
(Courtesy National Semiconductor Corp.)

214

8
Latches/Flip-Flops/Shift Registers

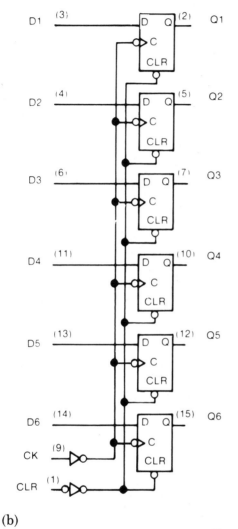

(b)

INPUTS		OUTPUTS		
CLEAR	CLOCK	D	Q	$\bar{Q}$
L	X	X	L	H
H	↑	H	H	L
H	↑	L	L	H
H	L	X	Q_0	$\bar{Q}_0$

(c)

H = High Level (steady state)
L = Low Level (steady state)
X = Don't Care
↑ = Transistion from low to high level
Q_0 = The level of Q before the indicated steady-state input conditions were established.

Figure 8-12. (continued)

8-8 Shift Registers

A shift register is a digital storage device that stores its information in flip-flops, where all flip-flops will be clocked at the same time.

Shift registers come in many different forms, the major difference being how the data is stored and retrieved from the shift register. There are common abbreviations for the input and output functions of a shift register, and they are defined in the

215

Digital Circuits and Devices

following list:

SISO - serial input/serial output
SIPO - serial input/parallel output
PISO - parallel input/serial output
PIPO - parallel input/parallel output
SISPO - serial input/serial, parallel output
PISPO - parallel input/serial, parallel output
SPISO - serial, parallel input/serial output
SPIPO - serial, parallel input/parallel output
SPISPO - serial, parallel input/serial, parallel output

TYPES OF INPUTS

Serial input occurs when data is applied to the shift register one bit at a time. Each bit shifted into the shift register requires one clock pulse.

Parallel input occurs when data is applied to each internal flip-flop at the same time. This type of input will load all bits in the shift register with only one clock pulse.

TYPES OF OUTPUTS

Serial output occurs when data is taken from the shift register one bit at a time. Each bit shifted out of the shift register requires one clock pulse. Also note the serial output is usually taken from the most significant bit MSB.

Parallel output occurs when data is taken from each output of the internal flip-flops at the same time. It requires no clock pulse as long as the data is in the register.

TYPES OF SHIFTING

A shift right shift register (most common) is a shift register that shifts its data from the least significant bit, LSB, to the MSB with the serial output being the MSB. The first bit shifted in the register must be the MSB.

A shift left shift register is a shift register that shifts its data from the MSB to the LSB with the serial output being the LSB. The first bit shifted in the register must be the LSB.

Shift registers contain flip-flops that are connected as data flip-flops. The simplest type of shift register is the SISO shift right shift register.

4-Bit SISPO Shift Right Shift Register

The 4-bit SISPO shift right shift register contains four data flip-flops connected in series with each other. Note the internal diagram shown in **Figure 8-13**.

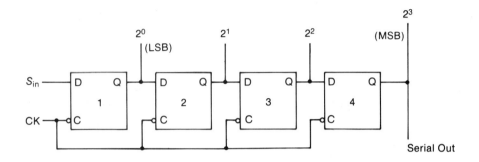

Figure 8-13 4-bit SISPO shift right shift register internal diagram

It should be noted that there is one data input called serial in (S_{in}). Data that is to be stored in the shift register must be applied beginning with the MSB and ending with the LSB.

SERIAL INPUT OPERATION

Each time new data is applied to the serial input and a clock pulse is received, the data on the serial input is transferred into flip-flop 1.

When the next bit of data is applied to the serial input and a clock pulse is received, the data on the serial input is transferred into flip-flop 1. At the same time, the data from flip-flop 1 is now transferred into flip-flop 2.

When the next bit of data is applied to the serial input and a clock pulse is received, the data on the serial input is transferred into flip-flop 1. At the same time, the data from flip-flop 1 is now transferred into flip-flop 2. Also, the data from flip-flop 2 is now transferred into flip-flop 3.

When the next bit of data is applied to the serial input and a clock pulse is received, the data on the serial input is transferred into flip-flop 1. At the same time, the data on flip-flop 1 is transferred to flip-flop 2. Data on flip-flop 2 is now transferred to flip-flop 3, and data on flip-flop 3 is transferred to flip-flop 4.

At this time all four bits of data are stored in the shift register. The data will remain unchanged in the shift register until a clock pulse is received.

As can be seen from the explanation of inputting serial data, it takes four clock pulses to input four bits of serial data into a four bit shift register.

Digital Circuits and Devices

OUTPUTTING DATA

Parallel Data

Parallel data is taken from 2^0 through 2^3 outputs. Once data is in the shift register, the data will be on the parallel outputs without needing a clock pulse to retrieve it.

Serial Data

The serial data output is taken from the MSB (2^3) of the shift register. In order to output four bits of data, four new bits of data must be entered into the shift register, serially. In order to output four bits of serial data, four clock pulses are required. This can be done by shifting a new four bit word into the shift register.

Figure 8-14a is the truth table for the 4-bit SISPO shift right register. **Figure 8-14b** shows the timing diagram for the shift register. Timing diagrams are important in the understanding of how and when data is moved from one register to another.

(a)

CK	S_{in}	2^0	2^1	2^2	2^3	S_{out}
0	X	X	X	X	X	X
	0	0	X	X	X	X
	0	0	0	X	X	X
	1	1	0	0	X	X
	1	1	1	0	0	0

(b)

Figure 8-14 4-bit SISPO shift right shift register
a) truth table and b) timing diagram

8
Latches/Flip-Flops/Shift Registers

From the timing diagram you should note that data is shifted in the shift register on only the *high* to *low* transition of the clock.

Number Of Clock Pulses For Data Transfer:

Serial input - 4 clock pulses.
Parallel output - 0 clock pulses.
Serial output - 4 clock pulses.
Serial input/serial output - 8 clock pulses.
Serial input/parallel output - 4 clock pulses.

4-BIT SPISPO SHIFT RIGHT SHIFT REGISTER (7495)

The four bit SPISPO shift right shift register, as shown in **Figure 8-15,** contains four RS flip-flops which act as data flip-flops. Also included are five 2 to 1 multiplexers. The multiplexers act as switches which allow data to be stored in either a serial or parallel format. When the M (mode) control line is *low,* the data input will be serial. When the M (mode) control line is *high,* data from the parallel inputs A, B, C, and D are transferred into the flip-flop.

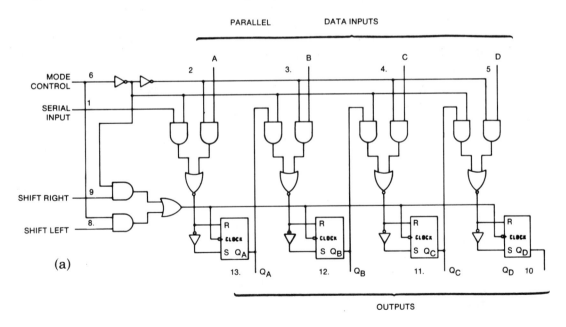

Figure 8-15 7495 a) logic diagram, b) connection diagram,
(Courtesy National Semiconductor Corp.)

Digital Circuits and Devices

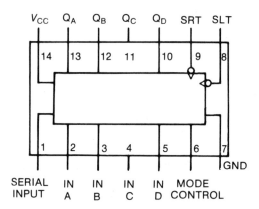

Figure 8-15. (continued)

As far as the serial load and the serial and parallel output operations are concerned, this shift register works the same as the SISPO shift register. To parallel load a four bit word into the shift register, the M control line must be *high* and a clock pulse must be applied to the SLT input. All parallel bits will be transferred into the shift register with one clock pulse.

It should also be noted that there are two clock inputs, SRT (shift right) and SLT (shift left). The SRT clock input shifts data to the right, while the SLT clock input moves data to the left. The mode control line must be *low* in order to shift data, and a high-to-low transition must occur on the corresponding clock input. As stated above, the SLT input is also used to load data into the shift register when the mode control is *high*.

Number Of Clock Pulses Required For Data Transfer:

Serial input - 4 clock pulses.
Parallel input - 1 clock pulse.
Serial output - 4 clock pulses.
Parallel output - 0 clock pulses.
Serial input/serial output - 8 clock pulses.
Serial input/parallel output - 4 clock pulses.
Parallel input/serial output - 5 clock pulses.
Parallel input/parallel output - 1 clock pulse.

WRITE/RECIRCULATE SHIFT REGISTER

The write/recirculate shift register function is a configuration, as shown in **Figure 8-16,** that allows data to be stored serially, when the mode line is in the write mode (*high*). When the mode line is in the recirculate mode (*low*), data is shifted right and the MSB is shifted back into the serial input. Data is never lost in this configuration. Instead, data is recirculated in the register.

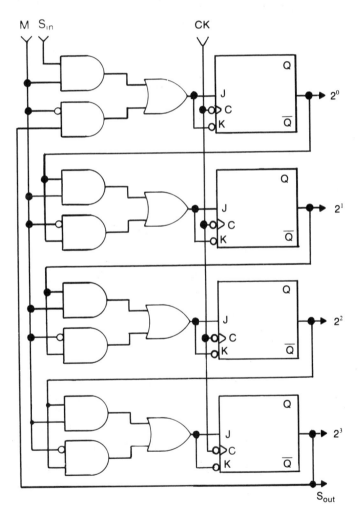

Figure 8-16 4-bit write/recirculate shift register with truth table

Digital Circuits and Devices

CK	M	D_{in}	2^0	2^1	2^2	2^3	
0	X	X	X	X	X	X	
↓	1	0	0	X	X	X	
↓	1	0	0	0	X	X	
↓	1	1	1	0	0	X	
↓	1	1	1	1	0	0	LOADING
↓	0	X	0	1	1	0	RECIRCULATING
↓	0	X	0	0	1	1	
↓	0	X	1	0	0	1	
↓	0	X	1	1	0	0	

Figure 8-16. (continued)

SHIFT REGISTER COUNTERS

Ring Counter

A ring or sequence counter is a shift register which allows only one output to be *high* at any one time. A 4-bit ring counter has a count of 0001, 0010, 0100, 1000 and then repeats itself. This type of counter is used to control digital circuits when the circuits must be turned on or off, one at a time. The ring counter must be preset before the count can first start. The preset condition is only required when the circuit is first powered up. A typical ring counter is shown in **Figure 8-17a**, with its truth table in **8-17b** and, its timing diagram in **8-17c**.

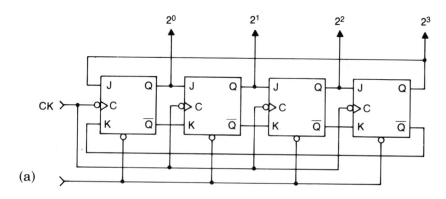

Figure 8-17 4-bit ring counter a) internal diagram, b) truth table, and c) timing diagram

8
Latches/Flip-Flops/Shift Registers

(b)

Clock	2^3	2^2	2^1	2^0
0	0	0	0	1
1	0	0	1	0
2	0	1	0	0
3	1	0	0	0

(c)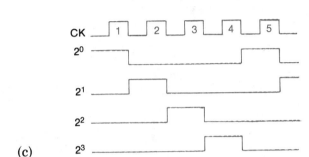

Figure 8-17. (continued)

Johnson Counter

A Johnson counter is a counter that will set each bit in a sequential order, beginning with the LSB. When all bits have been set, the counter will begin resetting each bit beginning with the LSB until all bits have been reset, then the sequence starts again. This counter also requires a start condition where all bits are reset when first powered up. Johnson counters are used for digital timing delays which in turn can control other digital devices. **Figure 8-18** shows: a) the internal diagram of a 5 bit Johnson counter; b) its truth table; and finally, c) its timing diagram.

(a)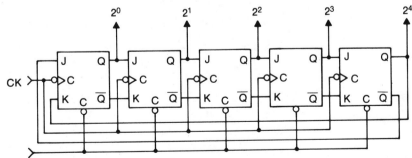

Figure 8-18. 5-bit Johnson counter a) internal diagram, b) truth table, and c) timing diagram

Digital Circuits and Devices

(b)
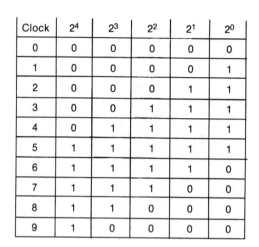

(c)
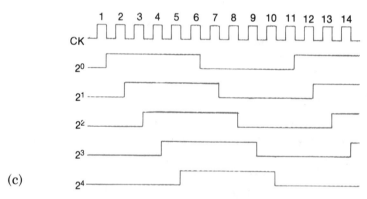

Figure 8-18. (continued)

ADDITIONAL USES FOR SHIFT REGISTERS

Since you can load data into a shift register in parallel and retrieve it serially, a shift register can be used as a parallel to serial converter. This function allows a computer to convert its parallel data to serial data.

Shift registers can also take serial data and convert it into parallel data to allow computers to receive data from a serial format.

Topic Review   8-8

1. A SISO shift register uses _____ input and _____ output formatting.

2. A flip-flop is used inside a shift register to store _____.

3. A PISO shift register _____ be used to convert parallel data into serial data.

4. When the serial and parallel clocks operate at the same frequency, an 8-bit binary word can be transferred the fastest using a _____ input and _____ output format.

Answers:

1. serial, serial
2. data
3. can
4. parallel, parallel

8-9 Summary Points

1. A latch is an electronic device which can store one bit of binary data.

2. In the modes of operation for latches and flip-flops, the names given to the operations are determined by the Q output.

3. The RS NAND latch operates in the opposite fashion as do any other type of latch or flip-flop.

4. When a latch or flip-flop is set, Q is *high*.

5. When a latch or flip-flop is reset, Q is *low*.

6. When a latch or flip-flop is in the *hold mode*, the outputs will not change from their previous levels.

7. When a latch or flip-flop is in the *disallowed mode*, the outputs of the device will be the same, which is a non-logical operation and should never occur.

8. When a latch or flip-flop is in the *disabled mode*, the outputs of the device will not change, no matter what the input levels are.

Digital Circuits and Devices

9. A latch is a level triggered device.

10. A flip-flop is an edge triggered device.

11. The internal device in a shift register that stores the data is a flip-flop which acts as a data flip-flop.

12. A shift register is a device that can store more than one bit of binary data.

13. Shift registers can act as a serial to parallel or parallel to serial converter.

14. Serial data is stored or retrieved one bit at a time.

15. Parallel data is stored or retrieved more than one bit at a time.

8-10 Chapter Progress Evaluation

1. What will be the level of $\overline{Q}$ when an RS latch is in the set mode?

2. What mode is an RS NOR latch in when R = O and S = O?

3. What mode is an RS NAND latch in when $\overline{R}$ = 1 and $\overline{S}$ = 1?

4. What mode will a clocked RS latch be in when R = O and S = 1?

5. What mode is a JK flip-flop in when J is *high*, K is *high*, and the proper clock pulse is received?

6. SISPO means what in reference to shift registers?

7. How many clock pulses does it take to shift an 8-bit word into a shift register in parallel format?

8. To input 4 bits of data serially and output the data in a parallel format, how many clock pulses does it take?

9. The type of shift register where data is not lost when shifting occurs is called?

10. Which type of device allows data to change on its output only once when the proper clock pulse is received?

Chapter 9

Counters

Objectives

Upon completion of this chapter, you should be able to do the following:

- Describe the operation and design of any type of digital counter

- Troubleshoot any type of digital counter, using either a truth table or timing diagram

- List the advantages and disadvantages of all types of counters in this chapter

9-1 Introduction

Counters are very useful digital devices and are found in many circuit designs today. These fundamental building blocks come in two basic constructions: *asynchronous* and *synchronous*. Both types will be studied in this chapter. They are also called sequential counters, and they can be designed to count up, down, and up/down. Nonsequential counters will also be investigated. They are commonly referred to as pseudo-random counters.

9-2 Standard Asynchronous Counters

In general, a *standard asynchronous counter* is an electronic device that counts in sequential order from zero to some power of two in binary. The *modulus* or *mod* of a counter is simply the number of times a counter increments before the counter repeats itself. Because asynchronous means serial or sequential, the output of the least significant section is used to clock the next section. The second sections's output is used to clock the input to the next section and so on, depending on the size of the counter. Typical values for *standard* counters are mod 2, 4, 8, and 16, or any other power of two. Normally the largest size counter on one IC is a mod 16, due to the type of internal circuitry needed for the counter and the maximum number of available

Digital Circuits and Devices

pins. Counters are made up from JK flip-flops. The number of flip-flops required to make a counter is based on a factor of two. Since the output of each flip-flop can represent two different conditions, it takes one flip-flop to make a mod two, two flip-flops to make a mod four, and so on.

MOD 2 COUNTERS

The *mod 2 counter* is one type of counter than can act as an up, down, up/down, up/down, asynchronous and synchronous counter without changing the way it is connected in the circuit. The reasons for this is that a mod 2 counter has only two states (0 and 1), and it contains only one flip-flop. The mod 2 counter is also known as a divide-by-two circuit. This means that the input clock frequency is divided by two.

Figure 9-1a shows the logic diagram for the mod 2 counter. Although the J and K inputs are not connected to anything on the drawing, these inputs of the flip-flop must be tied to supply or a logic high. Since in the real circuit J and K are tied *high*, the JK is in the toggle mode; this means that every time a clock pulse is received, the Q output of the JK will switch logic levels.

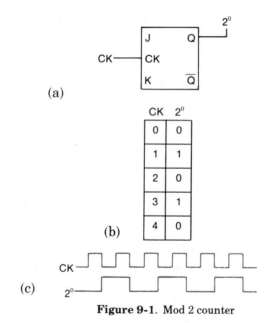

Figure 9-1. Mod 2 counter

As can be seen from the truth table of **Figure 9-1b**, the output (2^0) counts from a 0 to 1 and then restarts the count over again. **Figure 9-1c** shows its timing diagram. Since there are only two valid counts, it is known as a mod 2 counter. It should also

9 Counters

be noted that the *output* of the circuit is always taken from the Q output of the flip-flop and never $\overline{Q}$.

Since the counter counts in binary, the LSB is always labeled 2^0, and then increases by one for each additional output bit. From the timing diagram note the frequency from 2^0 is one half of the input clock frequency. This is why the circuit is also called a divide-by-two circuit.

ASYNCHRONOUS MOD 4 COUNTERS

Up Counter

An *asynchronous mod 4 up counter* is shown in **Figure 9-2a** with its truth table in **Figure 9-2b**. It counts in sequential order from 00 to 11 in binary totaling four different counts. Since it is an asynchronous counter, the output of FF-1 (flip-flop) is used as the clock input for FF-2 (flip-flop 2). The MSB is taken from the flip-flop that was clocked last.

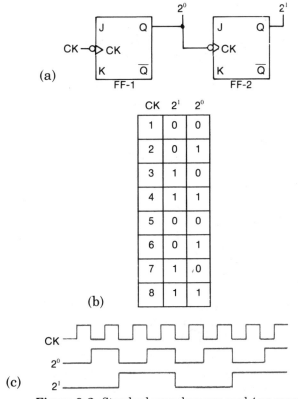

Figure 9-2. Standard asynchronous mod 4 up counter

Digital Circuits and Devices

Let's consider the *operation* of the asynchronous mod 4 up counter. In all cases refer to **Figure 9-2** which shows the logic diagram, truth table, and timing diagram for the asynchronous mod 4 up counter.

Clock 1:
 We will assume that when power is first applied to the circuit, that $2^0 = 0$ and $2^1 = 0$. (Count = 00).

Clock 2:
 On the falling edge of the clock pulse to FF-1, the Q output toggles from a *low* to a *high* ($2^0 = 1$). The *low* to *high* transition on the output of FF-1 is not seen as a proper clock input for FF-2. Since there was no clock pulse for FF-2, the output will not change and FF-2 is disabled ($2^1 = 0$). (Count = 01).

Clock 3:
 On the falling edge of the clock pulse to FF-1, the Q output toggles from a *high* to a *low* ($2^0 = 0$). The *high* to *low* transition on the output of FF-1 is seen as a proper clock input for FF-2. Since there was a proper clock pulse for FF-2, the Q output will toggle from a *low* to a *high* ($2^1 = 1$). (Count = 10).

Clock 4:
 On the falling edge of the clock pulse to FF-1, the Q output toggles from a *low* to a *high* ($2^0 = 1$). The *low* to *high* transition on the output of FF-1 is not seen as a proper clock input for FF-2. Since there was no clock pulse for FF-2, the output will not change, and FF-2 is disabled ($2^1 = 1$).(Count = 11). NOTE: At this time the counter is at its highest count.

Clock 5:
 On the falling edge of the clock pulse to FF-1, the Q output toggles from a *high* to a *low* ($2^0 = 0$). The *high* to *low* transition on the output of FF-1 is seen as a proper clock input for FF-2. Since there was a proper clock pulse for FF-2, the Q output will toggle from a *high* to a *low* ($2^1 = 0$). (Count = 00). The counter is now reset and restarts the count.

Remember that a mod 4 up counter counts from 00 to 11 in binary and then restarts the count.

Down Counter

The *asynchronous mod 4 down counter* shown in **Figure 9-3**, counts in sequential order backwards from 11 to 00 in binary totaling four different counts. Since it is an asynchronous down counter, the $\overline{Q}$ output from FF-1 is used as the clock input for FF-2. The MSB is taken from the flip-flop that was clocked last. Note the output for the counter in the logic diagram is still taken from the Q output from each flip-flop.

Let's consider the operation of the asynchronous mod 4 down counter. In all cases, refer to **Figure 9-3**.

Counters

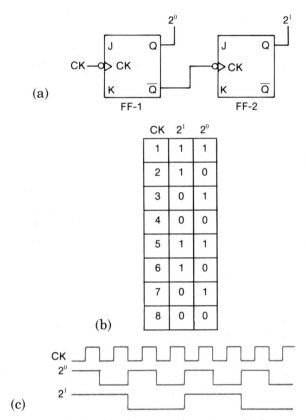

Figure 9-3. Standard asynchronous mod 4 down counter

Clock 1:
We will assume that when power is first applied to the circuit, that $2^1 = 1$ and $2^0 = 1$. (Count = 11).

Clock 2:
On the falling edge of the clock pulse to FF-1, Q and $\overline{Q}$ toggle. Q toggles from a *high* to a *low* ($2^0 = 0$). $\overline{Q}$ toggles from a *low* to a *high*. The *low* to *high* transition on output $\overline{Q}$ is applied to the clock input for FF-2. A *low* to *high* transistion is not a proper clock input for FF-2; therefore, FF-2 is now disabled and the output will not change. The Q output of FF-2 will remain <u>high</u> ($2^1 = 1$). (Count = 10).

Clock 3:
On the falling edge of the clock pulse to FF-1, Q and $\overline{Q}$ toggle. Q toggles from a *low* to *high* ($2^0 = 1$). $\overline{Q}$ toggles from a *high* to a *low*. The *high* to *low* transition on output Q is applied to the clock input for FF-2. A *high* to *low* transition is a proper clock input for FF-2; therefore, FF-2 will toggle. The Q output will change from

231

Digital Circuits and Devices

a *high* to a *low* ($2^1 = 0$). (Count = 01).

Clock 4:
On the falling edge of the clock pulse to FF-1, Q and $\overline{Q}$ toggle. Q toggles from a *high* to a *low* ($2^0 = 0$). $\overline{Q}$ toggles from a *low* to a *high*. The *low* to *high* transition on output $\overline{Q}$ is applied to the clock input for FF-2. A *low* to *high* transition is not a proper clock input for FF-2; therefore, FF-2 will be disabled, and the output will not change ($2^1 = 0$). (Count = 00). At this time the counter is at its lowest count.

Clock 5:
On the falling edge of the clock pulse to FF-1, Q and $\overline{Q}$ toggle. Q toggles from a *low* to a *high* ($2^0 = 1$). $\overline{Q}$ toggles from a *high* to *low*. The *high* to *low* transition on output $\overline{Q}$ is applied to the clock input for FF-2. A *high* to *low* transition is a proper clock input for FF-2; therefore, FF-2 will toggle. The Q output will change from a *low* to a *high* ($2^1 = 1$). (Count = 11). The counter now restarts the count.

Up/Down Counter

An *asynchronous mod 4 up/down counter* as shown in **Figure 9-4** counts in sequential order from either 00 to 11 in binary or 11 to 00 in binary, depending on whether the up/down control line is *high* or *low*. A 2 to 1 multiplexer is used to control whether the Q or $\overline{Q}$ output of FF-1 will be used for the clock input of FF-2. In the logic diagram of **Figure 9-4a**, a *high* applied to the U/D (up/down) control line enables AND gate 1 and disables AND gate 2. This allows the Q output from FF-1 to clock FF-2 and the counter will count up. When a *low* is applied to the U/D pin, AND gate 1 will be disabled, and AND gate 2 will be enabled. This allows the $\overline{Q}$ output from FF-1 to clock FF-2, and the counter will count down. For an explanation of the operation of the up/down counter when the U/D pin is *high*, see the explanation for the mod 4 up counter. See the explanation for the operation of the mod 4 down counter when the U/D pin is *low*.

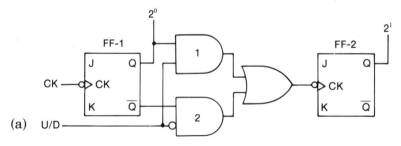

Figure 9-4. Standard asynchronous mod 4 up/down counter

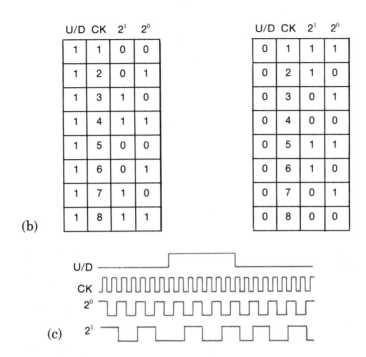

As can be seen from the timing diagram of **Figure 9-4c**, if the U/D line changes during a count, the count will begin counting in the corresponding direction on the next clock pulse.

LARGER ASYNCHRONOUS COUNTERS

In order to increase the count of an async mod counter, all that has to be done is to take the output of the last FF and use that output as the clock input for the new FF. This allows the count to double.

Mod 8 Up, Down, and Up/Down Counters

Figures 9-5, 9-6, and **9-7** illustrate the standard asynchronous mod 8 counters for the up, down, and up/down types, respectively. The operation of each will be the same as in the mod 4 counter with the exception of having more counts per counter.

When the U/D line is *low* for the up/down type of **Figure 9-7**, use the truth table for the standard async mod 8 down counter. When the U/D line is *high*, use the truth table for the standard async mod 8 up counter.

As can be seen from these examples, to increase the count simply add another JK flip-flop to the most significant bit (flip-flop). Use either Q or $\overline{Q}$ as the clock input to the last flip-flop, depending on if the counter will count up or down.

Digital Circuits and Devices

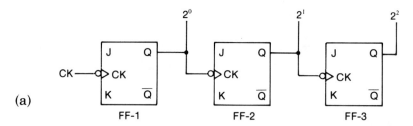

(a)

Figure 9-5. Standard asynchronous mod 8 up counter

(b)

CK	2^2	2^1	2^0
1	0	0	0
2	0	0	1
3	0	1	0
4	0	1	1
5	1	0	0
6	1	0	1
7	1	1	0
8	1	1	1
9	0	0	0

(b)

CK	2^2	2^1	2^0
1	1	1	1
2	1	1	0
3	1	0	1
4	1	0	0
5	0	1	1
6	0	1	0
7	0	0	1
8	0	0	0
9	1	1	1

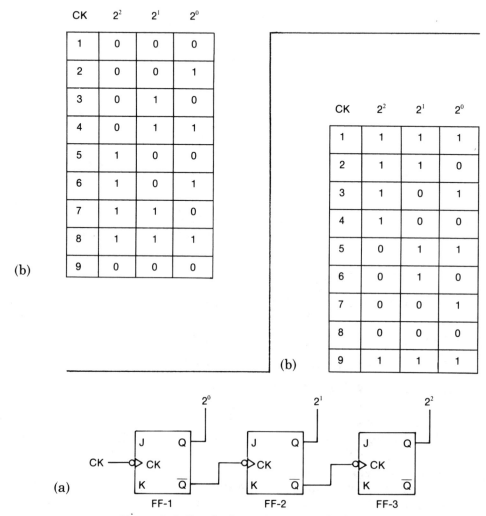

(a)

Figure 9-6. Standard asynchronous mod 8 down counter

9
Counters

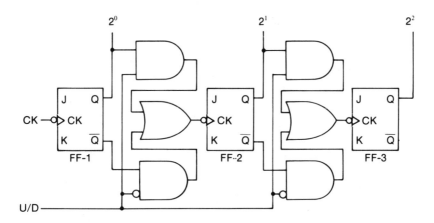

Figure 9-7. Standard asynchronous mod 8 up/down counter

Since the output of one flip-flop is used as a clock to the next flip-flop, the delay in the output change from the clock input to the output must be taken into consideration. The delays caused by the flip-flops are added together, and as the counter becomes larger, the delay becomes larger. This delay will limit the maximum frequency of operation; therefore, the majority of counters available in chip form are *synchronous* counters or a combination of asynchronous and synchronous counters. The asynchronous counter is also known as a *ripple counter*, because the output of each flip-flop is used to clock the clock input of the next flip-flop.

TTL 7493 Binary 4-Bit Ripple Counter

One type of ripple counter that is available today is the TTL 7493 binary four-bit ripple counter shown in **Figure 9-8**. This counter contains four JK flip-flops. Internally, the counter is set up as a mod 2 and mod 8. Each counter can be used separately, because both counters have their own clock inputs. This counter will only count up. There are two clock inputs labeled input A and B. Clock input A is the clock input for the mod 2, and clock input B is the clock input for the mod 8 counter. To make a mod 16 from the 7493, take the output from the mod 2 counter and use it as the clock input for the mod 8. In the case of a mod 16, Q^A is the LSB and Q^D is the MSB. Also included in the 7493 are two reset inputs, which will reset both counters at the same time and is an active *high* for the reset function. Notice that **Figure 9-8b** shows the count sequence truth table, while part **9-8c** is the reset/count truth table.

Digital Circuits and Devices

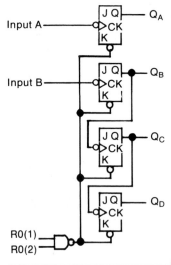

COUNT	OUTPUT			
	Q_D	Q_C	Q_B	Q_A
0	L	L	L	L
1	L	L	L	H
2	L	L	H	L
3	L	L	H	H
4	L	H	L	L
5	L	H	L	H
6	L	H	H	L
7	L	H	H	H
8	H	L	L	L
9	H	L	L	H
10	H	L	H	L
11	H	L	H	H
12	H	H	L	L
13	H	H	L	H
14	H	H	H	L
15	H	H	H	H

RESET INPUTS		OUTPUT			
R0(1)	R0(2)	Q_D	Q_C	Q_B	Q_A
H	H	L	L	L	L
L	X	COUNT			
X	L	COUNT			

Figure 9-8. 7493 Binary four-bit ripple counter

Topic Review 9-2

1. A _____ _____ counter can act as an asynchronous counter or a synchronous counter without changing how it is connected in the cirucit.

2. An _____ _____ counter uses the output of one JK flip-flop to clock the input of the next flip-flop.

3. A counter that counts from binary 1111 to 0000 is called a _____ _____ _____ counter.

4. A _____ _____ counter requires five JK flip-flops.

5. The two disadvantages of an _____ _____ counter are the increase in time delay as the counter becomes larger and a limited maximum operating frequency.

Answers:

1. mod 2
2. asynchronous mod
3. mod 16 down
4. mod 32
5. asynchronous mod

9-3 Standard Synchronous Counters

In general, **a standard synchronous mod counter is an electronic device that counts in sequential order from zero to some power of two in binary**. Because synchronous means at the same time, the clock pulse is applied to each flip-flop at *exactly the same time*. Typical values for a standard counter are 2, 4, 8, 16 or any other power of two. Because the clock is applied to each flip-flop at the same time, the time delay on a synchronous counter does not increase with the number of counts. In fact the time delay of a synchronous counter is the delay of the slowest flip-flop. Synchronous counters are the most popular because of their small time delay and ability to operate at high input clock frequencies.

The two most important modes of a JK flip-flop as far as a synchronous counter is concerned are the hold and toggle modes. The only problem with synchronous counters is if the counter is larger than a mod 4, an additional gate must be used to achieve the proper binary count.

Since the mod 2 counter works as a sync or async counter, we will move right into the synchronous mod 4 counter.

Digital Circuits and Devices

SYNCHRONOUS MOD 4 COUNTERS

Up Counter

A *synchronous* mod 4 up counter counts in sequential order from 00 to 11 in binary. Since the counter is synchronous, the clock is applied to each flip-flop at the same time. **Figure 9-9a** shows the diagram of a mod 4 up counter, while part **9-9b** shows its sequential truth table.

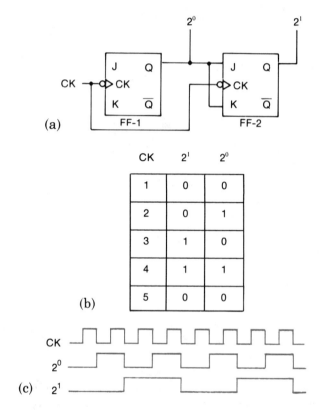

Figure 9-9. Standard synchronous mod 4 up counter

Note since each flip-flop is clocked exactly at the same time, the previous output level of one flip-flop must be used to determine how the next flip-flop will operate. It should also be noted that the least significant flip-flop is always in the toggle mode because J and K inputs are always *high*. The outputs of the first flip-flop will toggle each time a clock pulse is received.

Now let's consider the operation of the sync mod 4 up counter. For each case refer to **Figure 9-9**.

9 Counters

Clock 1:
We will assume that when power is first applied to the circuit, $2^0 = 0$ and $2^1 = 0$ (Count 00). At this time refer to **Figure 9-9a** and **c** for the following description of the mod 4 up counter.

Clock 2:
At the time of the clock, a *low* is applied to the J and K inputs of FF-2, from the Q output of FF-1. FF-2 is in the hold mode, so there is no change in its output ($2^1 = 0$). At the same time, the output of FF-1 will toggle from a *low* to a *high* ($2^0 = 1$). (Count 01).

Clock 3:
At the time of the clock, a *high* will be applied to the J and K inputs of FF-2 from the Q output of FF-1. The output of FF-2 will toggle from a *low* to a *high* ($2^1 = 1$). At the same time, the output of FF-1 will toggle from a *high* to a *low* ($2^0 = 0$). (Count 10).

Clock 4:
At the time of the clock, a *low* will be applied to the J and K inputs of FF-2, from the Q output of FF-1. The output of FF-2 will not change ($2^1 = 1$). At the same time, the output of FF-1 will toggle from a *low* to a *high* ($2^0 = 1$). (Count 11). **At this time the counter is at its highest count**.

Clock 5:
At the time of the clock, a *high* will be applied to the J and K inputs of FF-2, from the Q output of FF-1. The output of FF-2 will toggle from a *high* to a *low* ($2^1 = 0$). At the same time, the output of FF-1 will toggle from a *high* to a *low* ($2^0 = 1$). (Count 00). The counter is now reset and restarts the count.

Down Counter

The *synchronous mod 4 down counter* shown in **Figure 9-10**, counts in sequential order down from 11 to 00 in binary. Since it is a synchronous down counter, the $\overline{Q}$ output from FF-1 is used as the J and K inputs for FF-2. The outputs of the counter are still taken from the Q output from each flip-flop.

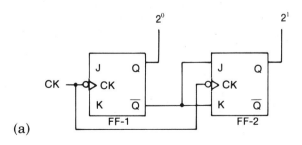

Figure 9-10. Standard synchronous mod 4 down counter

Digital Circuits and Devices

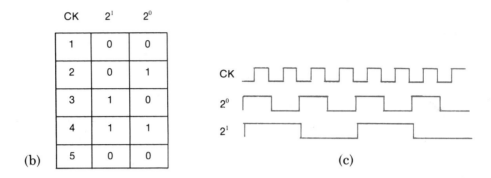

(b)   (c)

Figure 9-10. (continued)

Refer to **Figure 9-10** for the following description of the operation for mod 4 down counter.

Clock 1:
We will assume that when power is first applied to the circuit, $2^0 = 1$ and $2^1 = 1$. (Count 11).

Clock 2:
At the time of the clock, a *low* is applied to the J and K inputs of FF-2, from the $\overline{Q}$ output of FF-1. The output of FF-2 will not change ($2^1 = 1$). *At the same time*, the output of FF-1 will toggle from a *high* to a *low* ($2^0 = 0$). (Count 10).

Clock 3:
At the time of the clock, a *high* will be applied to the J and K inputs of FF-2 from the $\overline{Q}$ output of FF-1. The output of FF-2 will toggle from a *high* to a *low* ($2^1 = 0$). At the same time, the output of FF-1 will toggle from a *low* to a *high* ($2^0 = 1$). (Count 01).

Clock 4:
At the time of the clock, a *low* is applied to the J and K inputs of FF-2 from the $\overline{Q}$ output of FF-1. The output of FF-2 will not change ($2^1 = 0$). *At the same time*, the output of FF-1 will toggle from a *high* to a *low* ($2^0 = 0$). (Count 00). At this time the counter is at its lowest count.

Clock 5:
At the time of the clock, a *high* is applied to the J and K inputs of FF-2 from the $\overline{Q}$ output of FF-1. The output of FF-2 will toggle from a *low* to a *high* ($2^1 = 1$). *At the same time*, the output of FF-1 will toggle from a *low* to a *high* ($2^0 = 1$). (Count 11). The counter now restarts the count.

Up/Down Counter

The *synchronous mod 4 up/down counter* shown in **Figure 9-11** counts in a sequential order from either 00 to 11 in binary or 11 to 00 in binary. The direction of the count depends on whether the up/down control line (U/D) is *high* or *low*. A 2 to 1 multiplexer is used to control whether the Q or $\overline{Q}$ outputs of FF-1 will be applied to the J and K inputs of FF-2. In **Figure 9-11a**, when there is a *high* applied to the U/D control line, AND gate 1 will be enabled, while AND gate 2 will be disabled. This allows the Q output from FF-1 to be applied to the J and K inputs of FF-2 providing the counter to count up. When a *low* is applied to the U/D control line, AND gate 1 will be disabled, while AND gate 2 will be enabled. This allows the $\overline{Q}$ output from FF-1 to apply its level to the J and K inputs of FF-2, providing the counter to count down. For an explanation of the operation of the up/down counter see the explanation for the mod 4 up counter when the U/D line is *high*. Also, see the explanation for the mod 4 down counter when the U/D line is *low*.

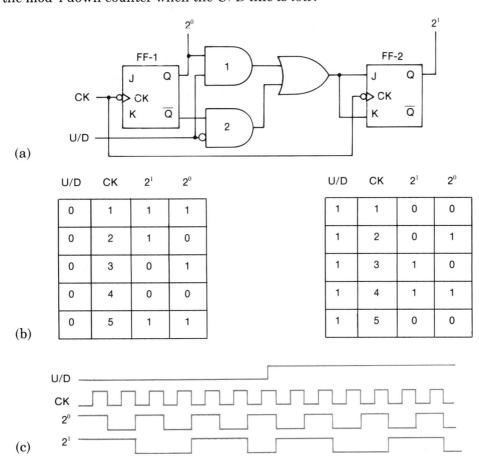

Figure 9-11. Standard synchronous mod 4 up/down counter

Digital Circuits and Devices

LARGER SYNCHRONOUS COUNTERS

Increasing the number of counts of a synchronous counter beyond a mod 4 is not as simple as in an asynchronous counter.

Mod 8 Up Counter

Figure 9-12a is a logic diagram of a *standard synchronous mod 8 up counter*. Note the connections of the AND gate that controls the level on the J and K inputs of FF-3.

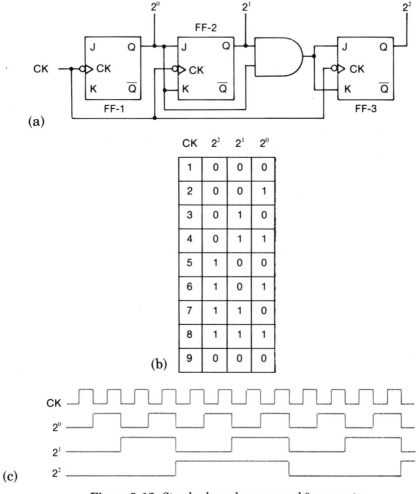

Figure 9-12. Standard synchronous mod 8 up counter

242

9
Counters

Refer to **Figure 9-12** for the following description of the mod 8 up counter.

Clock 1:
We will assume that when power is first applied to the circuit $2^0 = 0$, $2^1 = 0$, and $2^2 = 0$. (Count 000).

Clock 2:
At the time of the clock, a *low* is applied to the J and K inputs of FF-3, from the AND gate. The AND gate has a *low* output because the two inputs from the Q of FF-1 and FF-2 are both *low*. FF-3 is now in the hold mode, with no change in its output ($2^2 = 0$). *At the same time*, a *low* is applied to the J and K inputs of FF-2 from the Q output of FF-1. FF-2 is in the hold mode, with no change in its output ($2^1 = 0$). *Also at the same time*, the output of FF-1 will toggle from a *low* to a *high* ($2^0 = 1$). (Count 001).

Clock 3:
At the time of the clock, a *low* is applied to the J and K inputs of FF-3, from the output of the AND gate. The AND gate has a *low* output because FF-1 is *high* and FF-2 is *low*. FF-3 is now in the hold mode, with no change in its output ($2^2 = 0$). *At the same time*, a *high* is applied to the J and K inputs of FF-2, from the Q output of FF-1. FF-2 is in the toggle mode and its output will change from a *low* to a *high* ($2^1 = 1$). *Also at the same time*, the output of FF-1 will toggle from a *high* to a *low* ($2^0 = 0$). (Count 010).

Clock 4:
At the time of the clock, a *low* is applied to the J and K inputs of FF-3 from the output of the AND gate. The AND gate has a *low* output because FF-1 is *low* and FF-2 is *high*. FF-3 is now in the hold mode, so the output will not change ($2^2 = 0$). *At the same time*, a *low* is applied to the J and K inputs of FF-2 from the Q output of FF-1. FF-2 is in the hold mode, so there is no change in its output ($2^1 = 1$). *Also at the same time*, the output of FF-1 will toggle from a *low* to a *high* ($2^0 = 1$). (Count 011). The count is now sequential.

Clock 5:
At the time of the clock, a *high* is applied to the J and K inputs of FF-3, from the output of the AND gate. The AND gate has a *high* output because FF-1 is *high* and FF-2 is *high*. FF-3 is now in the toggle mode and the output of FF-3 will change from a *low* to a *high* ($2^2 = 1$). *At the same time*, a *high* is applied to the J and K inputs of FF-2 from the Q output of FF-1. FF-2 will change from a *high* to a *low* ($2^1 = 0$). *Also at the same time*, the output of FF-1 will toggle from a *high* to a *low* ($2^0 = 0$). (Count 100).

Clock 6:
At the time of the clock, a *low* is applied to the J and K inputs of FF-3 from the output of the AND gate. The AND gate has a *low* output because FF-1 is *low* and FF-2 is *low*. FF-3 is now in the hold mode, so there is no change in its output ($2^2 = 1$). *At*

Digital Circuits and Devices

the same time, a *low* is applied to the J and K inputs of FF-2 from the Q output of FF-1. FF-2 is in the hold mode, so its output will not change ($2^1 = 0$). *Also at the same time*, the output of FF-1 will toggle from a *low* to a *high* ($2^0 = 1$). (Count 101).

Clock 7:

At the time of the clock, a *low* is applied to the J and K inputs of FF-3 from the output of the AND gate. The AND gate has a *low* output because FF-1 is *high* and FF-2 is *low*. FF-3 is now in the hold mode, so there is no change in its output ($2^2 = 1$). *At the same time*, a *high* is applied to the J and K inputs of FF-2 from the Q output of FF-1. FF-2 is in the toggle mode, so its output will change from a *low* to a *high* ($2^1 = 1$). *Also at the same time* the output of FF-1 will toggle from a *high* to a *low* ($2^0 = 0$). (Count 110).

Clock 8:

At the time of the clock, a *low* is applied to the J and K inputs of FF-3 from the output of the AND gate. The AND gate has a *low* output because FF-1 is *low* and FF-2 is *high*. FF-3 is now in the hold mode, so there is no change in its output ($2^2 =$ EN1). *At the same time*, a *low* is applied to the J and K inputs of FF-2 from the Q output of FF-1. FF-2 is in the hold mode, so its output will not change ($2^1 = 1$). *Also at the same time* the output of FF-1 will toggle from a *low* to a *high* ($2^0 = 1$). (Count 111). This is the highest count for a mod 8.

Clock 9:

At the time of the clock, a *high* is applied to the J and K inputs of FF-3 from the output of the AND gate. The AND gate has a *high* output because FF-1 is *high* and FF-2 is *high*. FF-3 is now in the toggle mode, so the output will change from a *low* to a *high* ($2^2 = 0$). *At the same time*, a *high* is applied to the J and K inputs of FF-2 from the Q output of FF-1. FF-2 is in the toggle mode, so its output will change from a *high* to a *low* ($2^1 = 0$). *Also at the same time*, the output of FF-1 will toggle from a *high* to a *low* ($2^0 = 0$). (Count 000). At this time the counter starts the count again.

Mod 8 Down Counter

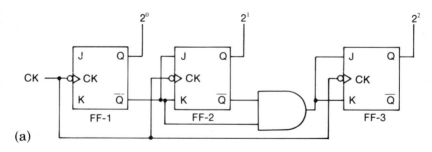

Figure 9-13. Standard synchronous mod 8 down counter

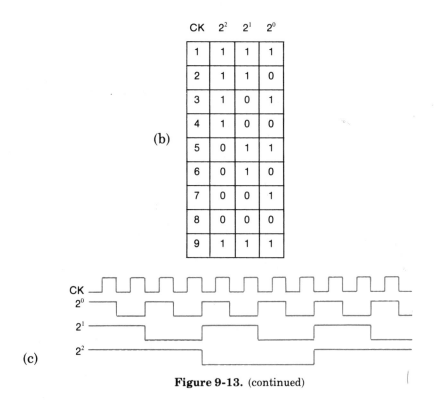

Figure 9-13. (continued)

A *standard sync mod 8 down counter* will count in a sequential order from a binary 111 to 000. In order for the counter to count down, the $\overline{Q}$ outputs from the flip-flops will be used as the J and K inputs of the next flip-flops. Since this is a synchronous counter, the AND gate must be used between the second and third flip-flop just like in the synchronous up counter, except that the $\overline{Q}$ outputs are used instead of the Q outputs. **Figure 9-13** is a diagram of a synchronous mod 8 down counter. The steps are the same as the synchronous up counter except the $\overline{Q}$ output is used as the input for the flip-flops.

Mod 8 Up/Down Counter

The *standard synchronous mod 8 up/down counter* shown in **Figure 9-14** will count up from 000 to 111 in binary or count down from 111 to 000 in binary, depending on the level on the U/D line. The counter will act as either an up or down counter. It uses multiplexers to control whether the flip-flop will use its Q or $\overline{Q}$ output to control the next flip-flop. Since it is a synchronous counter, the AND gate that controls the third flip-flop must be included in the circuit. Note in **Figure 9-14** the inputs to the AND gate and also how its output is connected in the counter circuit.

Digital Circuits and Devices

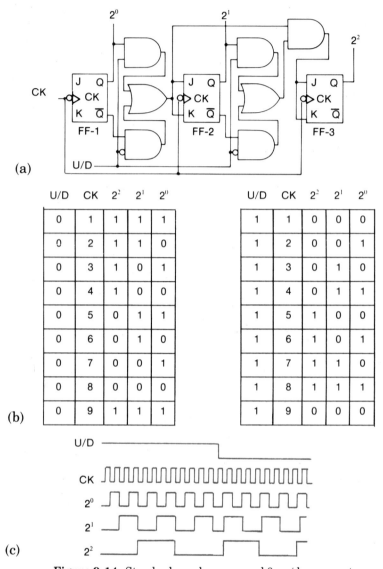

Figure 9-14. Standard synchronous mod 8 up/down counter

Mod 16 Up, Down, and Up/Down Counters

Parts (a), (b), and (c) of **Figure 9-15** illustrate the up, down, and up/down versions, respectively. Note the connections of the AND gates for each of these *standard sync mod 16 counters*. In general, to increase the count of a synchronous counter, simply add another JK flip-flop and another AND gate.

9
Counters

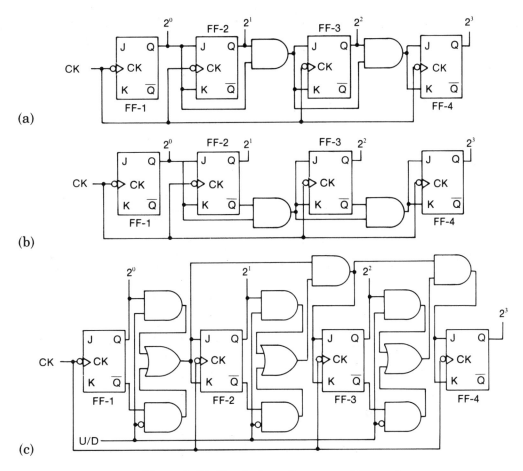

Figure 9-15. Standard synchronous mod 16 counters

TTL 74193 4-Bit Up/Down Counter

The 74193 is a synchronous four-bit (mod 16) up/down counter with parallel load inputs allowing the user to load a count into the counter. Shown in **Figure 9-16** are the logic and timing diagrams of the 74193. There are two clocks for this circuit. One is used as an up clock and the other is used as a down clock. Also included is a clear pin which will reset the counter being an active *high* input. The carry and borrow outputs are used when cascading more than one counter. The carry output will pulse *low* when the count of the counter is a binary 1111. This is used when counting up to trigger the next counter circuit. The borrow output will pulse *low* when the count of the counter is a binary 0000. This is used when counting down to trigger the next counter circuit.

247

Digital Circuits and Devices

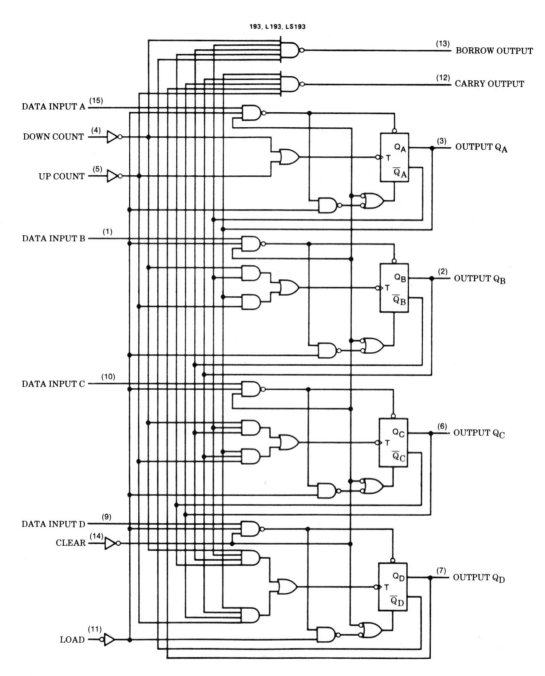

Figure 9-16. The 74193 synchronous four-bit up/down counter
Courtesy of National Semiconductor Corporation

9
Counters

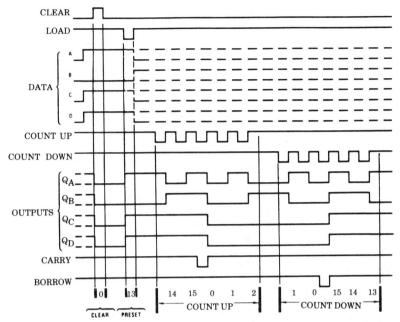

Sequence:

(1) Clear outputs to zero
(2) Load (preset) to binary thirteen
(3) Count up to fourteen, fifteen, carry, zero, one, and two
(4) Count down to one, zero, borrow, fifteen, and thirteen

Notes:

(A) Clear overrides load, data, and cuont inputs
(B) When counting up, count-down input must be high; when counting down, count-up input must be high

Figure 9-16. (continued)

To conclude, the advantages of using a synchronous counter is the short time delay and the ability to operate at a higher frequency than the asynchronous counter. The only disadvantage is it requires more circuitry to build, but since most come in chip form, this disadvantage is not important.

Topic Review 9-3

1. A _____ counter clocks all flip-flops at the same time.

2. The _____ mod 16 counter contains fewer components than the _____ mod 16 counter.

Digital Circuits and Devices

3. A mod 64 up counter will count to a maximum binary number of _____.

4. There are _____ 2 to 1 multiplexers in a mod 16 up/down counter.

5. Binary _____ will be the next count of a mod 16 down counter after the count of a binary of 0000.

Answers:

1. synchronous
2. asynchronous, synchronous
3. 111111
4. three
5. 1111

9-4 Non-Standard Synchronous Counters

A *non-standard synchronous counter* is a counter that contains a total number of counts per cycle that is other than a power of two. Examples of a non-standard mod counters are a mod 3, 5, 6, 7, 9, 10, 11 and so on. In simple terms, any counter that is not a power of two is called a non-standard mod counter. There are two basic ways of designing a non-standard mod counter. One is called the *forced reset method*. This is not normally used because of timing problems. Therefore, we will not go into any detail about this method. The second method is called the *cause/stop method*. Since the cause/stop method is the one used most of the time, we will go into great detail about using this method.

FORCED RESET METHOD

The *forced reset method* of designing a non-standard counter decodes the last count of the newly design counter and will force reset each flip-flop inside the counter on the next clock pulse. The next largest size standard mod counter must be used. It is then modified to reset the counter on the last count of new counter. **Figure 9-17** shows a mod 3 non-standard counter derived from the forced reset method along with its corresponding truth table.

Note that the Q output FF-2 will only go *high* on the binary count of two. This is the highest count in a mod 3 counter. The *high* along with the clock will cause the output of the NAND gate to reset the flip-flops. The reset function in a flip-flop forces the flip-flop to reset against any type of input condition and a clock signal. This requires a longer period of time than a standard input condition. Also, the normal level of the clock is *low*. When the clock goes *low* to *high* to *low* (clock pulse) and there is a binary count of two on the counter, the output of the NAND gate will go *low*. This will reset the flip-flops even though the normal inputs want the outputs to switch to different levels.

There are two disadvantages of the forced reset method. The first one is that it forces the flip-flops to clear from normal operation. This requires extra power to be used. The second disadvantage is that the clear circuitry usually operates at different time delays. The problem is that all flip-flops may not reset which prevents the entire counter from resetting. Normal counter operation will be especially hindered at higher frequencies.

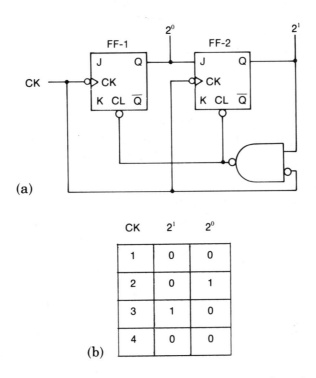

(a)

CK	2^1	2^0
1	0	0
2	0	1
3	1	0
4	0	0

(b)

Figure 9-17. Non-standard mod 3 counter derived from the forced reset method

CAUSE/STOP METHOD

The *cause/stop method* of designing non-standard mod counters uses normal operational input logic levels to cause the counter to reset. Since only input levels are changed there are no timing problems associated with this type of method. The following are the steps for designing a non-standard mod counter using the cause/stop method.

Digital Circuits and Devices

Design Procedure

1. Determine the value of the mod counter you wish to design.
2. Select the next largest standard synchronous mod counter. This counter will be modified to operate as the non-standard synchronous mod counter.
3. Determine the last count of the non-standard mod counter.
   last count = (non-standard mod # − 1)
4. Decode the last count so that there will be a *high* on only the last count. This will be called the *cause line*. If only one output will be *high* on the last count, do not use any decoding gate. If more than one output goes *high* at the last count, AND all of these outputs together.
5. Invert the decoded output with an inverter whose output will be used as the *stop line*. This line then will only go *low* at the last count.
6. Determine which flip-flop must be modified. Use the following four test outline to determine if any modification is needed.
   Test 1: If the output of the flip-flop is *high* on the last count and remains *high* on the next normal count, you must *cause* a reset to occur.
   Test 2: If the output of the flip-flop is *low* on the last count and changes to a *high* on the next normal count, you must *stop* the change from occurring.
   Test 3: If the output of the flip-flop is *low* on the last count and remains *low* on the next normal count, do not modify the flip-flop.
   Test 4: If the output of the flip-flop is *high* on the last count and *low* on the next normal count, do not modify the flip-flop.
7. After determining which flip-flop must be modified, use the following connections for the modification. If no flip-flop needs to be modified, then the design is complete.
   Cause: To *cause a change*, OR the cause line with only the K input of the flip-flop you wish to *cause* a change to occur.
   Stop: To *stop a change from occurring*, AND the stop line with only the J input of the flip-flop you wish to *stop* a change from occurring. If the first flip-flop must be stopped do not use an AND gate, just apply the *stop* line directly to its J input.

This method of making a non-standard counter may seem complex at first, but after some practice in applying these seven steps it will become easy.

Design and Operation of a Mod 3 Counter

In order to *design* a mod 3 counter using the cause/stop method, simply apply the seven steps previously outlined.

1. Type of counter mod 3.

9 Counters

2. Next largest standard synchronous mod counter is a mod 4 whose logic diagram was previously shown in **Figure 9-9**.
3. Last count is a binary 2. (Mod $3 - 1 = 2$).
4. Since the Q output of FF-2 will be the only *high* on the last count, the Q output of FF-2 will become the *cause line*.
5. Since the $\overline{Q}$ output will have the opposite condition on its output as compared to the Q output, the $\overline{Q}$ output of FF-2 will become the *stop line*.
6. Determine which flip-flop must be modified.

$$\begin{array}{lcc} & 2^1 & 2^0 \\ \text{Last count of a mod 3} = & 1 & 0 \\ \text{Next normal count} = & 1 & 1 \end{array}$$

Since 2^0 (output of FF-1) is *low* and will go to a *high* on the next normal count, the change must be *stopped* in FF-1. Since 2^1 (the output of FF-2) is *high* and will remain *high* on the next clock, we must *cause* a change to occur in FF-2.

7. Apply the *stop line* to the J input of FF-1. OR the *cause line* with the K input only of FF-2.

This results in the non-standard synchronous mod 3 up counter shown in **Figure 9-18**.

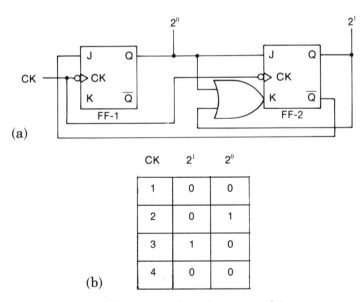

Figure 9-18. Non-standard synchronous mod 3 up counter

Referring to the truth table of **Figure 9-18b**, the operation of this counter is as follows.

Digital Circuits and Devices

Clock 1:
We will assume that when the first clock pulse is received, $2^0 = 0$, and $2^1 = 0$. (Count 00).

Clock 2:
At the time of the clock, a *low* is applied to the J input of FF-2 from the Q output of FF-1. Also since 2^1 is *low* and J is *low* the OR gate will apply a *low* to the K input of FF-2. FF-2 is in the hold mode, no change in its output ($2^1 = 0$). *At the same time*, there is a *high* applied to the J input of FF-1 from the $\overline{Q}$ output of FF-2. The K input of FF-1 is tied to supply. With input J and K of FF-1 *high*, the output of FF-1 will toggle from a *low* to a *high* ($2^0 = 1$). (Count 01).

Clock 3:
At the time of the clock, a *high* is applied to the J input of FF-2 from the Q output of FF-1. Also since 2^1 was *low* at the time of the clock and J is *high* at the time of the clock, the OR gate will apply a *high* to the K input of FF-2. FF-2 is in the toggle mode, and its output will change from a *low* to a *high* ($2^1 = 1$). *At the same time of the clock*, there is a *high* applied to the J input of FF-1 from the $\overline{Q}$ output of FF-2. The K input of FF-1 is tied to supply. With input J and K of FF-1 *high*, the output of FF-1 will toggle from a *high* to a *low* ($2^0 = 0$). (Count 10). The counter should not reset.

Clock 4:
At the time of the clock, a *low* is applied to the J input of FF-2 from the Q output of FF-1. Also since 2^1 is *high* and J is *low* at the time of the clock, the OR gate will apply a *high* to the K input of FF-2. Since FF-2 is now in the reset mode (J = 0, K = 1), its Q output will go *low* ($2^1 = 0$). *At the same time*, there is a *low* applied to the J input of FF-1 from the $\overline{Q}$ output of FF-2. The K input of FF-1 is tied to supply. FF-1 is now in the reset mode (J = 0, K = 1), the Q output will stay *low* ($2^0 = 0$). (Count 00). Now the circuit can start its count cycle again.

Design and Operation of a Mod 5 Counter

The following steps outline the design of a mod 5 counter using the cause/stop method.

1. Type of counter mod 5.
2. Next largest standard mod counter is a mod 8 whose logic diagram was previously shown in **Figure 9-12**.
3. Last count is a binary 4. (Mod 5 − 1 = 4).
4. Since the Q output of FF-3 will be *high* only at the last count, the Q output of FF-3 will become the *cause line*.
5. Since the $\overline{Q}$ output of FF-3 will have the opposite condition on it as compared to the Q output, the $\overline{Q}$ output will become the *stop line*.
6. Determine which flip-flop must be modified.

9
Counters

$$2^2 \; 2^1 \; 2^0$$

Last count of a mod 5   1   0   0

Next normal count   1   0   1

Since 2^0 (the output of FF-1) is *low* and will go *high* on the next clock pulse, this change must be *stopped* in FF-1.

Since 2^1 (the output of FF-2) is *low* and will stay *low* on the next clock pulse, no modification is required in FF-2.

Since 2^2 which is the output of FF-3 is *high* and will stay *high*, we must *cause* a change to occur in FF-3.

7. Apply the *stop line* to the J input of FF-1. OR the *cause line* with the K input only of FF-3.

This results in the non-standard synchronous mod 5 up counter shown in **Figure 9-19**.

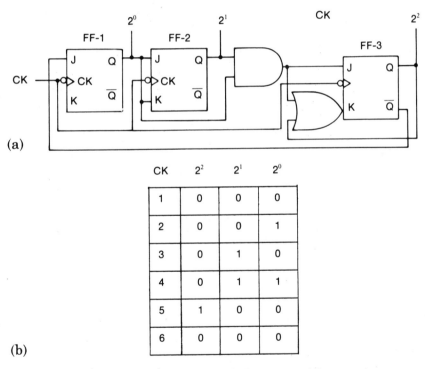

CK	2^2	2^1	2^0
1	0	0	0
2	0	0	1
3	0	1	0
4	0	1	1
5	1	0	0
6	0	0	0

(b)

Figure 9-19. Non-standard synchronous mod 5 up counter

Referring to the truth table of **Figure 9-19b**, the operation of this counter is as follows.

Digital Circuits and Devices

Clock 1:
We will assume when the first clock pulse is received $2^0 = 0$, $2^1 = 0$ and $2^2 = 0$. (Count 000).

Clock 2:
At the time of the clock, a *low* is applied to the J input of FF-3 from the AND gate being controlled by the Q outputs of FF-1 and FF-2. Since 2^2 is *low* and J is *low*, the OR gate will apply a *low* to the K input of FF-3. FF-3 is in the hold mode; therefore, no change occurs in its output ($2^2 = 0$). *At the same time*, there is a *low* applied to the J and K inputs of FF-2. FF-2 is in the hold mode with no change in its output ($2^1 = 0$). *Also at the same time*, there is a *high* applied to the J input of FF-1 from the Q output of FF-3. The K input of FF-1 is tied to supply. With input J and K of FF-1 *high*, the output of FF-1 will toggle from a *low* to a *high* ($2^0 = 1$). (Count 001).

Clock 3:
At the time of the clock, a *low* is applied to the J input of FF-3 from the AND gate off the Q outputs of FF-1 and FF-2. Since 2^2 was *low* at the time of the clock and J is *low* at the time of the clock, the OR gate will apply a *low* to the K input of FF-3. FF-3 is in the hold mode yielding no change in its output ($2^2 = 0$). *At the same time*, a *high* is applied to the J and K inputs of FF-2 from the Q output of FF-1. FF-2 is in the toggle mode. Its output will toggle from a *low* to a *high* ($2^1 = 1$). *Also at the same time*, there is a *high* applied to the J input of FF-1 from the Q output of FF-3. The K input of FF-1 is tied to supply. With input J and K of FF-1 *high*, the output of FF-1 will toggle from a *high* to a *low* ($2^0 = 0$). (Count 010).

Clock 4:
At the time of the clock, a *low* is applied to the J input of FF-3 from the AND gate. Since 2^2 was *low* at the time of the clock and J is *low* at the time of the clock, the OR gate will apply a *low* to the K input of FF-3. FF-3 is still in the hold mode. ($2^2 = 0$). *At the same time*, there is a *low* applied to the J and K inputs of FF-2 from the Q output of FF-1. FF-2 is in the hold mode, and its output will not change ($2^1 = 1$). *Also at the same time*, there is a *high* applied to the J input of FF-1 from the $\overline{Q}$ output of FF-3. The K input of FF-1 is tied to supply. With inputs J and K of FF-1 *high*, the output of FF-1 will toggle from a *low* to a *high* ($2^0 = 1$). (Count 011).

Clock 5:
At the time of the clock, a *high* is applied to the J input of FF-2 from the AND gate. Since 2^2 was *low* at the time of the clock and J is *high* at the time of the clock, the OR gate will apply a *high* to the K input of FF-3. FF-3 is in the toggle mode, and its output will toggle from a *low* to a *high* ($2^2 = 1$). *At the same time of the clock*, there is a *high* applied to the J and K inputs of FF-2 from the Q output of FF-1. FF-2 is in the toggle mode, and its output will toggle from a *high* to a *low* ($2^1 = 0$). *Also at the same time*, there is a *high* applied to the J input of FF-1 from the $\overline{Q}$ output of FF-3. The K input of FF-1 is tied to supply. With input J and K of FF-1 *high*, the output of FF-1 will toggle from a *high* to a *low* ($2^0 = 0$). (Count 100). The counter should reset

on the next clock.

Clock 6:

At the time of the clock, a *low* is applied to the J input of FF-3 from the AND gate. Since 2^2 is *high* at the time of the clock and J is *low* at the time of the clock, the OR gate will apply a *high* to the K input of FF-3. FF-3 is now in the reset mode (J = 0, K = 1), the Q output will go *low* (2^2 = 0). *At the same time of the clock*, there is a *low* applied to the J and K inputs of FF-2 from the Q output of FF-1. FF-2 is in the hold mode, and its output will not change (2^1 = 0). *Also at the same time*, there is a *low* applied to the J input of FF-1 from the $\overline{Q}$ output of FF-3. The K input of FF-1 is tied to supply. FF-1 is now in the reset mode (J = 0, K = 1), the Q output will go *low* (2^0 = 0). (Count 000). Now the circuit can start its count cycle again.

Mod 6, 7, 9, and 10 Counters

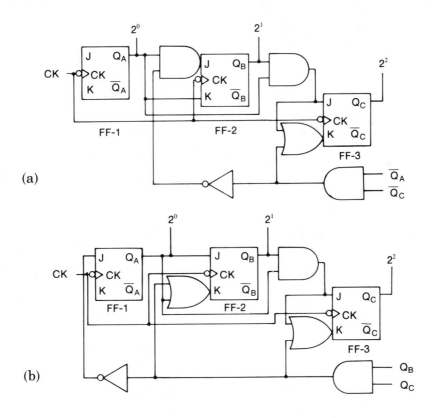

Figure 9-20. Non-standard synchronous mod 6 and mod 7 up counters

Digital Circuits and Devices

Figures 9-20 and **9-21** show more examples of non-standard mod counters. Note that modifications are again made to the standard mod counter to make the non-standard counters. Also each of these counters are of the synchronous type identified by the clock signal being applied to each flip-flop simultaneously.

The standard mod 8 up counter can also be modified to obtain the non-standard mod 6 up counter of **Figure 9-20a** and the non-standard mod 7 up counter of **9-20b**. **Figures 9-21a** and **9-21b** show the non-standard synchronous counters being mod 9 and mod 10, respectively. These two non-standard counters were derived from the standard synchronous mod 16 up counter.

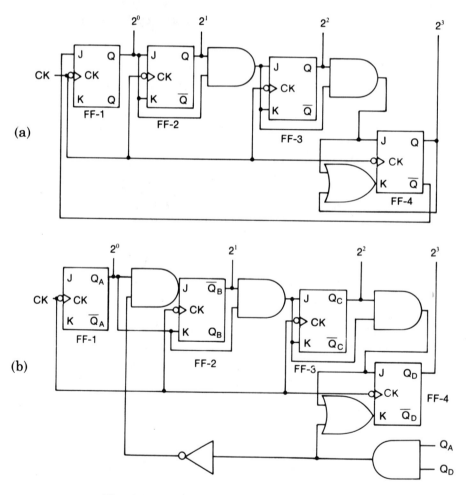

Figure 9-21. Non-standard synchronous mod 9 and mod 10 up counters

Counters

Mod 6 and Mod 10 by Cascading Counters

Another way of making a mod 6 counter is by applying the clock to a mod 2 counter and then taking the output of the mod 2 counter and using it as the clock input of the synchronous mod 3 counter. When combining the two types of counters the time delay will be twice as long as the synchronous mod 6 counter. The time delay of the mod 2 counter is in series with the time delay in the synchronous mod 3 counter.

Figure 9-22a shows this example of a non-standard mod 6 up counter by combining a standard mod 2 up counter and a non-standard synchronous mod 3 up counter.

This technique can be extended. For example, a non-standard mod 10 up counter, as shown in **Figure 9-22b**, is the result of cascading a non-standard synchronous mod 5 up counter to a standard mod 2 up counter.

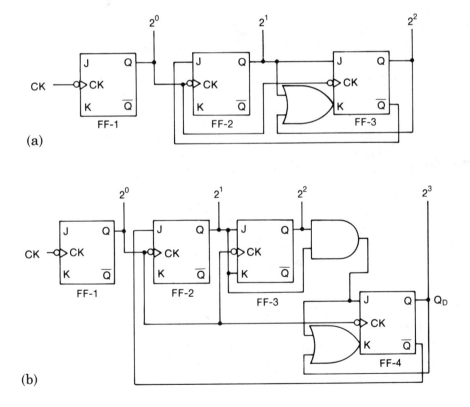

Figure 9-22. Cascading counters to obtain non-standard async/sync mod 6 and mod 10 up counters

Digital Circuits and Devices

TTL Integrated Counters

Two examples of TTL IC non-standard counters are the 74160 and the 74912. The 74160 is a synchronous mod 10 decode counter with a parallel load. This chip counts only in the up direction. The 74192 is a synchronous mod 10 up/down counter also with parallel load.

Topic Review 9-4

1. A _____ _____ counter is one that contains a total number of counts per cycle that is other than a power of two.

2. The two ways of designing a non-standard mod counter are by using either the _____ _____ method or the _____ method.

3. The major disadvantages of the _____ _____ _____ are extra power dissipation and slower toggle rate for normal counter operation.

4. There are no timing problems associated with the _____ _____.

5. Cascading a _____ counter and any _____ _____ counter can result in a larger non-standard synchronous counter.

Answers:

1. non-standard synchronous
2. forced reset, cause/stop
3. forced reset method
4. cause/stop method
5. mod 2, non-standard synchronous

9-5 Pseudo Random Counters

A *pseudo random counter* is a counter that counts in a non-sequential order from zero to some number. Pseudo means simulated; therefore, a pseudo random counter counts in a simulated random count. The designer can choose any order as long as no number is repeated in the count cycle.

DESIGN PROCEDURE

In order to design a pseudo random counter follow the steps given in the follow-

9 Counters

ing procedure.

1. Determine the count sequence.
2. Determine the number of sequentially clocked JK flip-flops required for the cirucit. Note, one flip-flop is required for each output bit of the counter.
3. Make a truth table using the output of the counter to determine the input condition needed at the input of each flip-flop to produce the output condition at the time of the next clock. For example, consider the truth table shown in **Table 9-1**. This table shows the input conditions needed on each input to cause the output to change to the proper level on the next clock.

Table 9-1

Output			Inputs	
Old Level		New Level	J	K
0	to	1	1	X
1	to	0	X	1
0	to	0	0	0
0	to	1	1	X

When X is a don't care condition, this means that it does not matter if the input is *low* or *high* for the change to occur.

4. From the truth table, K-map all *highs* and *lows* for each input. In all unfilled locations in the K-map, place an X (don't care). Loop all *highs* in the K-map to determine the circuit needed for the inputs. Don't care conditions can be looped with the *highs* to achieve the largest power of two in the loop. However, all don't care conditions do not have to be mapped.
5. Draw the resulting pseudorandom counter.

DESIGN EXAMPLE #1

Design a pseudo random counter that will count 00, 01, 11, 10, and then reset.

The Design

1. Count 00, 01, 11, 10 and reset.
2. Requires two JK flip-flops.
3. Truth table given in **Table 9-2**.
4. From this truth table, K-maps are derived to determine the circuits needed for the J and K inputs of each flip-flop. These K-maps are illustrated in **Figure 9-23** and outlined as follows:
   (a) shows the K-map for the J input of FF-1 yielding $J = Q_B$
   (b) shows the K-map for the K input of FF-1 yielding $K = Q_B$
   (c) shows the K-map for the J input of FF-2 yielding $J = Q_A$
   (d) shows the K-map for the K input of FF-2 yielding $K = Q_A$

Digital Circuits and Devices

Table 9-2
Truth table for design example #1

Outputs		FF-1 Inputs		FF-2 Inputs	
Q_B	Q_A	J	K	J	K
0	0	1	X	0	X
0	1	X	0	1	X
1	1	X	1	X	0
1	0	0	X	X	1
0	0	RESET		RESET	

(a) (b) (c) (d)

Figure 9-23. K-maps for design example #1

The resulting logic diagram of the pseudo random counter is shown in **Figure 9-24a**.

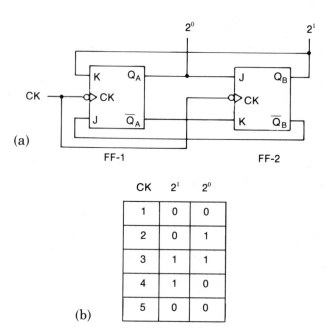

CK	2^1	2^0
1	0	0
2	0	1
3	1	1
4	1	0
5	0	0

Figure 9-24. Resulting pseudo random counter of design example #1

Counters

The Operation

Clock 1:
 We will assume on the first clock pulse that $2^0 = 0$ and $2^1 = 0$. (Count 00).

Clock 2:
 At the time of the clock, a *low* is applied to the J input of FF-2 from the Q_A output of FF-1. A *high* is applied to the K input of FF-2 from the $\overline{Q_A}$ output of FF-1. With J *low* and K *high* on FF-2, FF-2 is in the reset mode. The Q_B output of FF-2 will be *low* ($2^1 = 0$). *At the same time of the clock*, a *high* is applied to the J input of FF-1 from the $\overline{Q_B}$ output of FF-2. A *low* is applied to the K input of FF-1 from the Q_B of FF-2. With J *high* and K *low* on FF-1, FF-1 is in the set mode. The Q_A output of FF-1 then will be *high* ($2^0 = 1$). (Count 01).

Clock 3:
 At the time of the clock, a *high* is applied to the J input of FF-2 from the Q_A output of FF-1. A *low* is applied to the K input of FF-2 from the $\overline{Q_A}$ output of FF-1. With J *high* and K *low* on FF-2, FF-2 is in the set mode. The Q_B output of FF-2 will be *high* ($2^1 = 1$). *At the same time of the clock*, a *high* is applied to the J input of FF-1 from the $\overline{Q_B}$ output of FF-2. A *low* is applied to the K input of FF-1 from the Q_B output of FF-2. With J *high* and K *low* on FF-1, FF-1 is in the set mode. The Q_A output of FF-1 will be *high* ($2^0 = 1$). (Count 11).

Clock 4:
 At the time of the clock, a *high* is applied to the J input of FF-2 from the Q_A output of FF-1. A *low* is applied to the K input of FF-2 from the $\overline{Q_A}$ output of FF-1. With J *high* and K *low* on FF-2, FF-2 is in the set mode. The Q_B output of FF-2 will be *high* ($2^1 = 1$). *At the same time of the clock*, a *low* is applied to the J input of FF-1 from the $\overline{Q_B}$ output of FF-2. A *high* is applied to the K input of FF-1 from the Q_B output of FF-2. With J *low* and K *high* on FF-1, FF-1 is in the reset mode. The Q_A output of FF-1 will be *low* ($2^0 = 0$). (Count 10). This is the last count. On the next clock a reset should occur.

Clock 5:
 At the time of the clock, a *low* is applied to the J input of FF-2 from the Q_A output of FF-1. A *high* is applied to the K input of FF-2 from the $\overline{Q_A}$ output of FF-1. With J *low* and K *high* on FF-2, FF-2 is in the reset mode. The Q_B output of FF-2 will be *low* ($2^1 = 0$). *At the same time of the clock*, a *low* is applied to the J input of FF-1 from the $\overline{Q_B}$ output of FF-2. A *high* is applied to the K input of FF-1 from the Q_B output of FF-2. With J *low* and K *high* on FF-1, FF-1 is in the reset mode. The Q_A output of FF-1 will be *low* ($2^0 = 0$). (Count 00). The new clock cycle starts again. The truth table for this pseudo random counter is shown in **Figure 9-24b**.

DESIGN EXAMPLE #2

Design a pseudo random counter that will count 000, 111, 010, 101 and reset.

Digital Circuits and Devices

The Design

1. Count 000, 111, 010, 101 and reset.
2. Requires three JK flip-flops.
3. Truth table given in **Table 9-3**.

Table 9-3.
Truth table for design example #2

Outputs			FF-1 Inputs		FF-2 Inputs		FF-3 Inputs	
Q_C	Q_B	Q_A	J	K	J	K	J	K
0	0	0	1	X	1	X	1	X
1	1	1	X	1	X	0	X	1
0	1	1	1	X	X	1	1	X
1	0	1	X	1	0	X	X	1
0	0	0						

4. From this truth table, K-maps are derived to determine the circuits needed for the J and K inputs of each flip-flop. These K-maps are illustrated in **Figure 9-25** and outlined as follows:

   (a) shows the K-map for the J input of FF-1 yielding $J = 1$
   (b) shows the K-map for the K input of FF-1 yielding $K = 1$
   (c) shows the K-map for the J input of FF-2 yielding $J = \overline{Q_C}$ or $\overline{Q_A}$
   (d) shows the K-map for the K input of FF-2 yielding $K = \overline{Q_C}$ or $\overline{Q_A}$
   (e) shows the K-map for the J input of FF-3 yielding $J = 1$
   (f) shows the K-map for the K input of FF-3 yielding $K = 1$

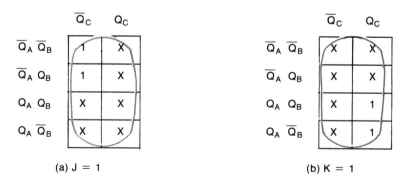

Figure 9-25. K-maps for design example #2

9
Counters

FF-2 K-maps

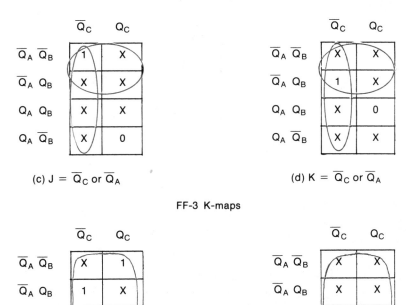

(c) $J = \overline{Q}_C$ or $\overline{Q}_A$

(d) $K = \overline{Q}_C$ or $\overline{Q}_A$

FF-3 K-maps

(e) $J = 1$

(f) $K = 1$

5. The resulting logic diagram of the pseudo random counter along with its truth table are shown in **Figure 9-26**.

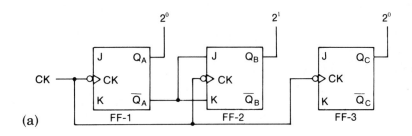

Figure 9-26. Resulting pseudo random counter of design example #2

265

Digital Circuits and Devices

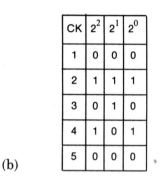

(b)

Figure 9-26. (continued)

As can be seen from the two examples, as the number of bits increases in a pseudo random counter, the complexity of design will also increase. For this reason, pseudo random counters are restricted to four output bits or less.

If a larger pseudo random counter is nedded, a RAM (Random Access Memory) chip is programmed to give the proper count. A standard counter will access the data from the RAM counter. The RAM counter can be any size with any count without increasing the size of the circuit to any great extent. The only problem is that it will have to be reprogrammed every time the power is first applied to the circuit. If reprogramming is not wanted in a memory counter, a ROM (Read Only Memory) chip can be used. This device will retain data and cannot be reprogrammed once the chip is made.

Topic Review 9-5

1. A _____ _____ _____ is a type of counter that counts in a non-sequential order.

2. The word pseudo means _____.

3. A pseudo random counter can count in any order as long as no number is _____ in the count cycle.

4. _____ flip-flops are required to make a pseudo random counter with a count of 0000, 0111, 1111 and then reset.

Answers:

1. pseudo random counter
2. simulated
3. repeated
4. Four

9-6 Summary Points

1. A standard mod counter has a total number of counts per cycle that is a power of two.

2. Standard and non-standard counters count in sequential order from zero to some value.

3. An asynchronous counter uses the output of one flip-flop as the clock input for the next flip-flop.

4. A standard asynchronous counter is an electronic device that counts in sequential order from zero to some power of two in binary.

5. The modulus or mod of a counter is simply the number of times a counter increments before the counter repeats itself.

6. The mod 2 counter is the only type of counter that can act as an up, down, up/down, asynchronous and synchronous without changing the way it is connected in the circuit.

7. A synchronous counter has all flip-flops clocked at the same time.

8. An up counter counts up from zero to its maximum count.

9. A down counter counts down from its maximum count to zero.

10. A non-standard counter has a total number of counts per cycle other than a power of two.

Digital Circuits and Devices

8. A pseudo random counter counts in a non-sequential order.

9-7 Chapter Progress Evaluation

1. Which size mod counter can be a sync or async counter without changing its connections in the circuit?

2. How many JK flip-flops are needed to build a mod 32 up counter?

3. How many AND gates are required to make a sync mod 4 up/down counter?

4. When a high is on the U/D line of the counter shown in **Figure 9-14a**, the counter will sequentially count in which direction?

5. What is the maximum binary count of a non-standard synchronous mod 6 up counter?

6. In order to design a mod 9 counter, you must start with what size counter?

7. Which flip-flops must be modified in a mod 12 counter? Consider FF-1 as the LSB (may have more than one answer).

8. From question 7, in which flip-flops must you *stop* a change from occurring? (may have more than one answer)

9. From question 7, in which flip-flops must you *cause* a change to occur? (may have more than one answer)

10. What is the maximum number of different counts a 3-bit pseudo random counter can have?

11. If a larger pseudo random counter is needed, what two types of chips can be utilized?

Chapter 10

Digital and Analog Conversion

Objectives

Upon completion of this chapter, you should be able to do the following:

* Determine the output voltage of any size WRN or R2R

* Determine the resolution of any type of DAC

* List the advantages and disadvantages of any type of DAC given in this chapter

* Describe the basic operation of any type of ADC given in this chapter

* List the advantages and disadvantages of any type of ADC given in this chapter

10-1 Introduction

Digital information cannot interface *directly* with analog information. Therefore, digital-to-analog (D/A) conversion is needed for a digital computer to send data to an analog device. Conversely, analog-to-digital (A/D) conversion prefaces data received from an analog device, so that it can be handled by a digital computer.

Since most industrial devices and almost all chemical or physical measurements are analog in nature, D/A and A/D methods of conversion are widely used. Digital-to-anaglog converters (DACs) are used in vector CRT (cathode-ray tube) display systems, audio synthesizers, automatic testing cells, controlled attenuators, and process controllers. In addition, most analog-to-digital converters (ADCs) contain a DAC. As a result of these various applications, it is essential that you understand the basic principles of D/A and A/D conversion.

10-2 Digital-to-Analog (D/A) Applications

Digital Circuits and Devices

Let's now go into greater detail with D/A applications. **Figure 10-1** illustrates the use of two DACs in a graphics display application. A computer controls the CRT display. The computer sends digital data to the Y-axis DAC, which converts it to a DC voltage and positions the CRT vertically. Likewise, the X-axis DAC converts digital data to a DC voltage that positions the beam horizontally. If each DAC handles eight bits, the screen can show 256 beam spots in each direction.

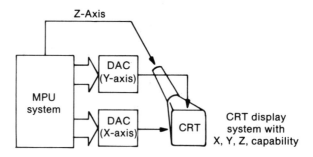

Figure 10-1. Graphic display application of DACs

After the computer has positioned the beam spot, it turns on the spot with a signal to the Z-axis. The Z-axis signal may be digital. However, adding a DAC to the Z-axis could convert this system, where its signal could vary the intensity of the spot instead of merely turning it off or on.

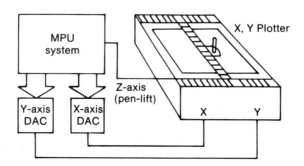

Figure 10-2. XY plotter using DAC control

A similar system, with a time delay added in the computer program, could drive an XY plotter as shown in **Figure 10-2**. The time delay allows the slower XY plotter to keep up with the computer's XY information. Z-axis signals control the up-and-down movement of the pen. A logic high lifts the pen, and a low sets it down. Not all XY plotters take analog input signals. Some drive the X and Y movements of the pen with stepper motors.

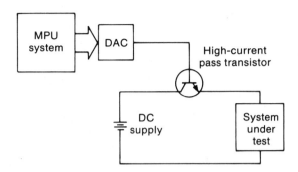

Figure 10-3. An automatic testing setup using a DAC

Figure 10-3 is a simplified diagram of part of an automatic testing setup. In this application, a computer controls the operating voltage for the system being tested. A DAC converts the digital computer output to an analog DC voltage, and applies it to the base of a high-current-pass transistor. The emitter voltage of the transistor equals base voltage minus $0.7V$. Hence, the computer controls the voltage applied to the system under test. An ADC attached to a computer input port (not shown) enables the computer to measure the voltage in the system under test.

Figure 10-4 shows the basics of an MPU-controlled audio or waveform synthesizer. Here, the computer generates digital words that a DAC converts into an analog voltage. The DAC applies the analog signals to an amplifier and speaker. Software programming will cause the computer to output the proper series of digital patterns to generate any waveform desired.

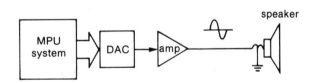

Figure 10-4. MPU-controlled waveform synthesizer

10-3 Digital-to-Analog Hardware

WRN

A popular DAC design is the *weighted-resistor-network* (WRN), shown in **Figure 10-5**. Notice two operational amplifiers are used. The first op amp (A1) is configured as an inverting summing amplifier. Each input (D_3, D_2, D_1, and D_0) operates its respective switch. When an input goes to a logic *high*, the corresponding switch is activated placing a $5\ V$ reference to its weighted resistor. Conversely, a logic *low* effectively connects the resistor to ground potential. Each of the switches shown

Digital Circuits and Devices

represents a transistor biased to operate in saturation.

The second op amp (A2) is configured as an inverting buffer amplifier. Therefore, this circuit converts four bits of digital data into a positive DC voltage proportional to the applied digital data.

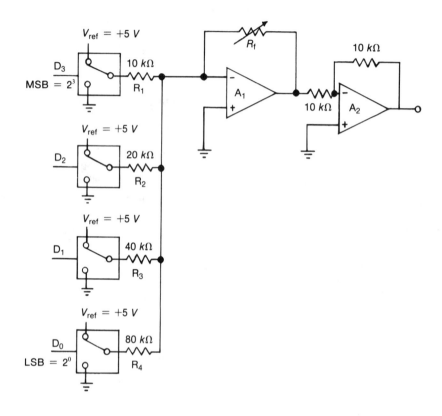

Figure 10-5. A DAC using a weighted-resistor network

To measure the analog output of this DAC, connect a voltmeter to measure the output voltage from A2. Adjust R_f to 1.6 kohm. This establishes A1 with a nominal gain of -3.5. Now, if each of the sixteen possible digital input combinations from 0000 to 1111 were applied, the voltmeter would read from 0 to $1.5\,V$ in sixteen $0.1\,V$ steps. Thus, the DAC converts the digital input into an analog signal with a resolution of $0.1\,V$.

From this we can define DAC *resolution* as the smallest change in the output voltage in a DAC for each binary count. **Table 10-1** shows the output voltage for each binary count of a four-bit input given that the DAC resolution is 0.1 V/step.

Table 10-1.
Output voltage versus binary count

Binary Count	Output Voltage (V)
0000	0
0001	0.1
0010	0.2
0011	0.3
0100	0.4
0101	0.5
0110	0.6
0111	0.7
1000	0.8
1001	0.9
1010	1.0
1011	1.1
1100	1.2
1101	1.3
1110	1.4
1111	1.5

Another way of determining the output of a weighted-resistor network is by using the *WRN formula*. The equation is:

$$V_{out}\text{WRN} = \left[\frac{V_{in}[2^3] + V_{in}[2^2] + V_{in}[2^1] + V_{in}[2^0]}{2^N - 1}\right] \times \left[\frac{R_f}{\text{TPR}}\right]$$

where: V_{in} is the voltage at each input
2^x is the numerical value for each input
N is equal to the total number of inputs in the DAC
TPR is the total parallel resistance of weighted input resistors

Example: What is the output voltage of the DAC in **Figure 10-5** given that $R_f = 1.6$ *k*ohm, the binary input is 1010, and a logic *high* is equal to $5V$?

first:
$$\text{TPR} = \cfrac{1}{\cfrac{1}{10k} + \cfrac{1}{20k} + \cfrac{1}{40k} + \cfrac{1}{80k}}$$

$$\text{TPR} = \frac{1}{.1 + .05 + .025 + .0125}$$

$$\text{TPR} = \frac{1}{0.1875} = 5.33 \; k\text{ohm}$$

next:
$$2^0 = 1, \; 2^1 = 2, \; 2^2 = 4, \; 2^3 = 8, \; 2^N = 2^4 = 16$$

therefore,

$$V_{out}\text{WRN} = \left[\frac{5[8] + 0[4] + 5[2] + 0[1]}{16 - 1}\right] \times \left[\frac{1.6k}{5.33k}\right]$$

$$V_{out}\text{WRN} = 1.0 \; V$$

As the number of input bits to a DAC increase, the amount of output voltage change per binary count will decrease. We call this an increase in resolution. If we were to add an additional input to the circuit in **Figure 10-5**, the MSB resistor would have a value of 5 $k$ohms. The resolution would increase from 0.1 V per count to 0.05 V per count. So by increasing the total number of inputs a DAC has, it can control external devices much more accurately.

If R_f were set to 16 $k\Omega$, the resolution becomes 1 V. If the op amp operates from a $-15\;V$ bipolar supply, its output variations beyond about 12 V would *not* be linear. This DAC would therefore provide 1 V resolution, from 0 V to about $-12\;V$, in 1 V steps. (The switches in the boxes are actually transistors that are turned on or off by the logic input).

The reason the transistors are used is that the outputs of digital devices do not output exactly the same *high* and *low* levels. If the DAC is to work properly the *high* and *low* levels to all binary inputs must be exactly the same, otherwise each output voltage step will not be equal.

There are two basic disadvantages to this circuit however: most of the resistors are not standard values, and the circuit becomes impractical if you attempt to increase resolution beyond eight bits.

Any time you convert a digital count into an analog voltage there will be some error in the conversion. Quantization refers to the approximate conversion of a digital to analog signal. Quantization error is therefore the difference between the theoretical and actual output voltage of a DAC. The more inputs to a DAC the smaller the quantization error.

R2R

A somewhat similar circuit, called an R2R ladder, uses only two resistor values, as shown in **Figure 10-6**. The circuit draws its name from the fact that one resistor value is twice that of the other. As in the weighted-resistor circuit, the R2R ladder connects between the digital source and an op amp. The op amp prevents load variations from being reflected back onto the ladder, and permits scaling of the output beyond the reference voltage value. That is, the maximum or top output voltage depends on the supply, not on the input reference voltage.

Again, let's assume a voltage reference at 5 V. Digital inputs, therefore, apply either 5 V (logic *high*) or 0 V (logic *low*) to each "rung" of the R2R ladder. Under these conditions, op amp ladder output ranges from 0 to 4.6875 V, in sixteen 0.3125 V steps, with 0 V being the first step, and 4.6875 V being the sixteenth step. The top value can be scaled up or down by varying R_f in the op amp circuit: for example, if you wanted the output to spread from 0 to +1.5 V, you would set R_f to 3.2 $k\Omega$.

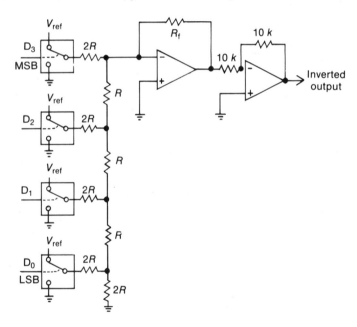

Figure 10-6. R2R binary ladder

The output impedance of an R2R ladder is always equal to the value of R, no matter what binary count is on the inputs. To find the output voltage of the R2R ladder in **Figure 10-6**, when $R = 10\ k$ and $2R = 20\ kohms$, take the voltage at the MSB and divide it by two, add it to the voltage at 2 and divide it by four, add it to the

Digital Circuits and Devices

voltage at 1 and divide it by 8, add it to the voltage at LSB and divide it by 16. Then take the sum, multiply it times R_f and divide by the value of R.

$$V(\text{R2R}) = \left[\frac{V \text{ at MSB}}{2} + \frac{V \text{ at } 2^1}{4} + \frac{V \text{ at } 2^2}{8} + \frac{V \text{ at LSB}}{16}\right] \times \left[\frac{R_f}{R}\right]$$

Always divide the MSB voltage by two and then increase the division of each additional input bit by a factor of two, until all bits have been accounted for.

Example 1: Find the output voltage of the 4-bit R2R ladder in **Figure 10-6**, with a binary input of 0111 and an R_f of 10 kohm and the value of $R = 10$ kohms. A logic high is equal to 5 V.

$$V(\text{R2R}) = \left[\frac{0}{2} + \frac{5}{4} + \frac{5}{8} + \frac{5}{16}\right] \times \left[\frac{10k}{10k}\right] = 2.1875 \ V$$

Example 2: Find the output voltage of the 4-bit R2R ladder in **Figure 10-6**, with a binary input of 0111 and an R_f of 10 kohm and the value of $R = 20$ kohms. A logic high is equal to 5 V.

$$V(\text{R2R}) = \left[\frac{0}{2} + \frac{5}{4} + \frac{5}{8} + \frac{5}{16}\right] \times \left[\frac{10k}{20k}\right] = 1.09375 \ V$$

As in the WRN, if the number of inputs to an R2R ladder is increased, the output voltage per count will decrease, and it is said that the resolution of the DAC has increased.

10-4 Testing a DAC

There are two ways of testing the operation of a DAC: one is called the static test, and the other is called the monotonicity test, which is a dynamic test.

STATIC TEST

To perform the static test, place a binary number onto the digital inputs and check the output voltage to make sure the output voltage for each step is exactly the same level. If the voltage levels are not the same, it means that the values of the resistors are not exact.

MONOTONICITY TEST

This is a visual inspection of a DAC output to determine any malfunction. This test is performed by applying a counter circuit to the binary inputs of the DAC and then clocking the counter circuit at a high enough frequency to allow a visual display

10
Digital and Analog Conversion

on an oscilloscope. Since with every count of the counters the output voltage from the DAC will increase in equal voltage steps, the output of the DAC will look like a staircase. If there is no problem with the DAC, each step will be equal; if a resistor is missing or has changed values, the display of the staircase waveform will change. The following is a waveform for a good DAC. Also included are DACs with problems; the problems will be stated with each waveform.

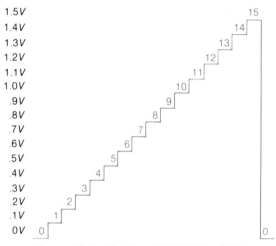

Figure 10-7. Working 4-bit WRN or R2R

Figure 10-7 shows the waveform of a good 4-bit WRN or R2R with a resolution of 1 V/count. Note that there are 16 steps starting with 0 through 15. Also note that each step is of equal height and width. If the heights are not equal, the value of the resistors are not equal.

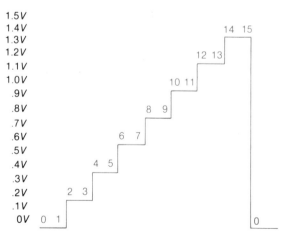

Figure 10-8. DAC with LSB resistor missing

Digital Circuits and Devices

In **Figure 10-8** the width of the steps are twice as wide as the good DAC and instead of each step increasing at a rate of 0.1 V per count, it is increasing at a rate of 0.2 V per count. This means that the resistor for the LSB, R_4 in **Figure 10-5** is missing. When R_4 is missing, each count that uses the LSB will be lower in voltage level.

Figures 10-9, 10-10 and **10-11** are a few additional examples of the DAC shown in **Figure 10-5** with various resistors removed.

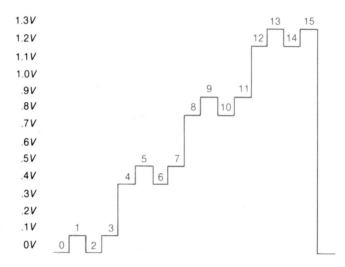

Figure 10-9. DAC with R_3 missing

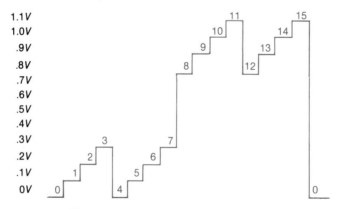

Figure 10-10. DAC with resistor R_2 missing

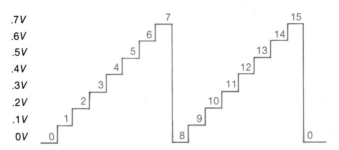

Figure 10-11. DAC with resistor R_1 missing

By examining the preceding examples it should be noted that when an input resistor is missing, any output voltage level that uses that resistor will fall to the voltage level minus that missing voltage level. In the case of a resistor changing its value to a low resistance, the output voltage that uses the resistance will decrease. If the resistor value increases, any voltage level that uses that resistor value will also increase. These types of problems usually only occur in discrete DACs. DACs that are built into integrated chips do not exhibit the same problems. Another advantage of a DAC in chip form is that it takes less board space.

Topic Review   10-2 through 10-4

1. A/An _____ is needed for a computer to control an external analog device.
2. A 4-bit R2R is _____ accurate than a 5-bit WRN.
3. _____ is the maximum number of steps in a 6-bit R2R ladder.
4. If resolution goes up in a WRN, the voltage per output step goes _____.
5. The two tests used when checking the operation of a DAC are _____ and _____.

Answers:

1. DAC
2. less
3. 64
4. down
5. static, monotonicity

10-5 Analog-to-Digital (A/D) Applications

It's now time to take a look at the opposite side of the conversion process: analog-

Digital Circuits and Devices

to-digital conversions (ADC). Such A/D conversion is required when a digital computer receives input data from an analog device. Typical A/D applications include automatic testing and measurement, digitizing of audio or video signals, and of course, industrial process control.

Figure 10-12a illustrates part of an automatic test setup. Here, the processor will measure digital voltage signals from a test node in the unit being tested. A single ADC, with an analog multiplexer driven by a processor, can perform voltage tests hundreds of times faster and more accurately than the sharpest technician.

In **Figure 10-12b**, another automatic testing setup is shown. Here, the setup measures DC current. The op amp inverts the voltage across a sense resistor. This op amp multiplies voltage from the resistor by a factor equal to the amplifier's closed-loop gain. (The voltage varies directly with current through the sense resistor). You can calibrate the voltage ouput of the op amp to indicate test-circuit current.

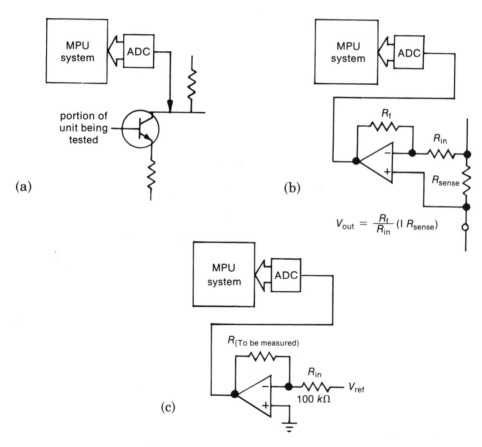

Figure 10-12. Automatic testing procedures

10
Digital and Analog Conversion

This technique can be used to measure resistance, as shown in **Figure 10-12c**. Input voltage to the op amp forms a fixed, accurate reference. The gain of the circuit is the variable, and is the ratio of the feedback resistor to the input resistor. That is:

$$A_v = \frac{R_f}{R_{in}}$$

in which:
A_v = op-amp gain
R_f = value of feedback resistor, in ohms
R_{in} = value of input resistor, in ohms

Since the input resistance is fixed, voltage output depends on the value of the resistor being tested.

Audio and video signals can be digitized using fast ADC components. **Figure 10-13** lays out the concept of digitizing audio waveforms; however, far more is involved than this simple sketch shows. The processor determines from the ADC when to load digital information into system memory. (Usually, special filters break the audio signal into frequency bands; then, each band forms a different digital word). Digitizing video is more complicated: the processor must keep up with the horizontal and vertical sync (synchronization) as well as the video.

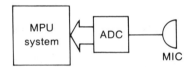

Figure 10-13. Digitizing audio waveforms

10-6 Analog-to-Digital Hardware

The most popular types of ADC are successive approximation, integration (or ramp), voltage to frequency, counter and servo, and parallel. These ADC types vary in conversion speed, accuracy, size, versatility, and cost; naturally, certain ADCs suit some applications better than others.

SUCCESSIVE-APPROXIMATION ADC

Figure 10-14 is a diagram of a successive-approximation (SA) type of ADC. Let's consider the principles of operation of this simple 4-bit converter.

Note first that the op amp is wired as a comparator: the open-loop, maximum-gain connection provides rapid switching of the output states, with but a few microvolts of input difference. One comparator input is the analog input V_{ANA} to be converted to a digital value. The other input is the equivalent of the digital value held by the output register.

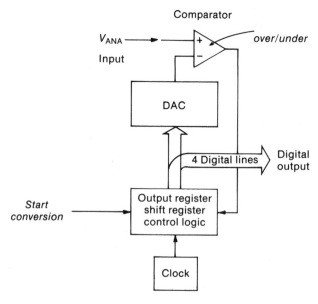

Figure 10-14. Successive-approximation ADC

Note also that a DAC is part of this ADC. The second input to the comparator comes from this DAC, which converts the digital voltage of the output register into an analog voltage. (Many ADCs contain internal DACs).

Figure 10-14 shows a shift register, control logic, and the output register as one block function. To understand the operation, assume an analog input V_{ANA} of 6.5 V which must be converted to digital form. First, the control logic receives a *start conversion* signal. It loads the left most (fourth) bit of a shift register with logic 1. Thus, the shift register initially contains the 4-bit binary pattern 1000.

This binary 1000 goes through the output register to the DAC. The DAC converts 1000 to its analog equivalent of 8 V, which is applied to the comparator. The comparator matches the 8 V input with the 6.5 V analog input and sends an *over* signal back to the control logic. In response to this *over* signal, the control logic discards the 1 in the most significant position of the shift and output registers. Hence, the comparator indicates that analog input is less than 8 V.

Now, the second most significant (third) bit takes on logic 1. The binary pattern 0100 goes to the DAC, which converts it to 4 V and sends it to the comparator. The comparator finds the 4 V and sends it to the comparator. The comparator finds the 4 V from the register smaller than the analog input of 6.5 V and sends an *under* signal to the control logic. In response, the control logic leaves the third bit set *high* and makes the next most significant (second) bit in the shift register a logic *high*. The binary number 0110 goes to the DAC. The DAC converts 0110 to the analog equivalent value of 6 V, which is fed to the comparator. The comparator matches the 6 V input from the register with the 6.5 V input and returns an *under* signal to the control logic.

Now the control logic keeps the third and second bits *high* and sets the first bit *high*, to form the binary number 0111. The control logic sends the number 0111 to the D/A converter, where it's converted to 7 V and forwarded to the comparator. The 7 V now exceeds the 6.5 V analog input, and an *over* signal is sent back to the control logic. The first bit is changed back from a *high* to a *low*, to form the binary number 0110. Conversion ends, since all four bits have successively been tried.

The output register now contains 0110 — the binary equivalent to decimal 6. This is as close as a 4-bit SA ADC can come to representing the actual analog input of 6 V without exceeding it. This 0110 pattern latches into the output register and becomes available as 4-bit parallel output data.

Conversion tolerance for this simple SA ADC stands essentially at one whole count. This means that even if the analog input voltage had been only a few microvolts under 7 V, this converter would still have indicated it as 6 V. Because of this very coarse resolution, a 4-bit converter is little used. However, much closer tolerance can be obtained by increasing the number of bits. Among the converters in popular use today, for example, resolution varies from 8 to 16 bits, with a 12-bit size common.

ADCs are widely used as interfaces with computers, since they offer both high resolution and speed. They also provide a fixed conversion time that is independent of the analog input voltage, so that one conversion doesn't affect another.

Analog Devices' ADC 1140 converter, offers 16-bit SA accuracy and speed. This kind of ADC is suitable for automatic test equipment, medical and nuclear instrumentation, seismic data acquisition, pulse-code modulation (PCM) telemetry, and industrial robotic applications.

INTEGRATION (OR RAMP-TYPE) ADC

Another type of ADC uses a voltage ramp. **Figure 10-15** shows a dual-ramp, or integration, type of converter. Upon receiving a *start conversion* signal, the control logic circuit does three things: first, it resets the binary up counter to 0000; second, it applies (and then removes) a short circuit around the integrator feedback capacitor, C; third and finally, it connects the input of the integrator op amp to the analog input voltage, V_{ANA}. (Of course, in a practical ADC, the switches are analog types: field-effect transistors, or FETs).

Digital Circuits and Devices

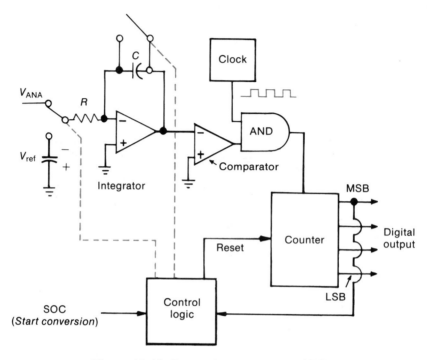

Figure 10-15. Integration or ramp-type ADC

A positive analog-input voltage produces a negative-going ramp at the integrator output; this negative slope signal feeds the negating input of the comparator. As soon as the ramp exceeds a few negative microvolts, output of the comparator goes *high*, enabling the AND gate. The ramp continues moving negative.

The AND gate, when enabled, presents clock pulses to the counter, where the pulses are counted. The counter runs all the way to 1111, after which the next clock pulse resets the counter to 0000. (If the counter is a BCD — binary-coded decimal type, as it is in some ADCs, the maximum count reaches only 1001 which is decimal 9, the highest BCD 4-bit number possible).

The instant the counter resets, the control logic flips the input switch, applying an accurate negative voltage reference to the negative input of the integrator. This starts the output ramp positive-going from its farthest negative point, thus reflecting the analog-input value.

When this second integrator ramp reaches 0 V, the comparator inhibits the AND gate. This freezes the counter at a count that is proportional to the time the ramp takes to drop to zero. With the system calibrated, this count indicates the digital value nearest to (but less than) the analog-input voltage.

Consider the concept again while studying **Figure 10-16**. Time t_1 represents the period during which the unknown analog input is applied to the integrator. The fixed value of R and C sets the time t_1. The analog input determines the steepness of the slope — that is, how far from zero it goes in time t_1. In other words, V_{ANA} controls the amount of negative-charge buildup across capacitor C.

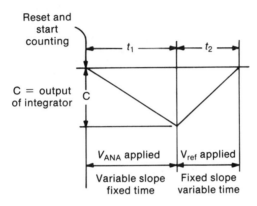

Figure 10-16. V_{ANA} and V_{ref} on single-ramp converter

The value of the reference voltage V_{REF} fixes the steepness of the discharge, or positive-going, slope. Therefore, time t_2 varies according to the magnitude of C, which corresponds to the analog input. During t_2, the counter accumulates a value in proportion to the analog input. At the end of t_2, with C discharged, the comparator switches the counter off.

A single-ramp converter integrates a reference voltage that is opposite in polarity to the analog signal. A counter tallies clock pulses until the integrator output equals analog input; this makes the count proportional to the ratio of the analog input to the reference input. (Accuracy depends on the capacitor and the clock frequency; however, single-ramp converters are seldom used).

A double-ramp converter offers several advantages over a single-ramp converter. Conversion accuracy depends on neither the capacitor value nor the clock frequency, because they affect both the slopes in an offsetting or negating manner. At the same time, a dual-slope ADC works too slowly for fast data acquisition. It's better suited for relatively slow analog signals, such as the outputs from thermocouples or gas chromatographs. Ditigal voltmeters, too, generally contain dual-slope ADCs. Also, errors in either the integrator or comparator op amp show up in the digital output. One company developed and patented a scheme, called the *quad-slope principle*, for nullifying such input errors. Their 12-bit AD7552, incorporates this principle. This ADC stores errors in the form of a digital count during a calibration cycle, and subtracts them from the final count during the conversion cycle.

Digital Circuits and Devices

VOLTAGE-TO-FREQUENCY (V/F) ADCS

An infrequently used method of A/D conversion translates an analog-input voltage into a pulse-train output. This ADC consists of a precision multivibrator that varies according to the analog-input voltage. The output of the multivibrator is a pulse train, the frequency of which is proportionally accurate to the analog-input signal.

The output of a V/F converter is serial — that is, a single signal varying in frequency. (Other ADCs deliver parallel outputs). A counting circuit must interpret the output of a V/F ADC. A computer can be used to count the number of pulses occurring in a given period; this count indicates frequency, thus also indicating analog input voltage.

V/F converters provide high resolution and long-term precision for two-wire data transmissions, with high noise immunity. Conversion, however, takes longer than in the other systems discussed. Conversions can occur as fast as 10 per second, but the resolution will improve as measurement times are increased.

COUNTER AND SERVO ADCS

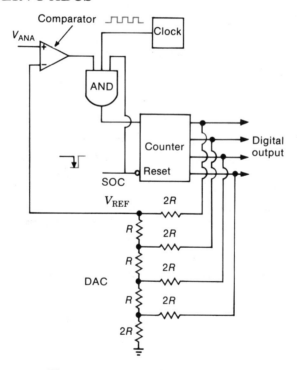

Figure 10-17. 4-bit counter ADC

A 4-bit counter type of ADC appears in **Figure 10-17**. Upon receiving a momentary start-of-conversion (SOC) pulse (a logic *low*), the counter resets to 0000. Besides providing parallel digital output, the counter feeds an R2R DAC. (You probably recall that the successive approximation converter also contained a DAC).

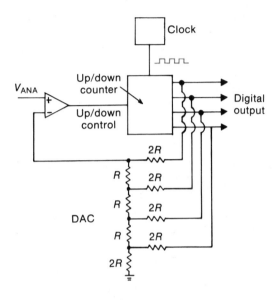

Figure 10-18. Servo ADC

The DAC converts counter output, initially 0000, into a corresponding analog voltage and applies it to a comparator. The analog voltage to be converted is the other comparator input. Comparator output stays at logic *high* as long as analog input exceeds the DAC value brought from the counter.

Output of the AND gate follows the clock signal, since the other two AND inputs (after the SOC pulse) are *high*. Thus, the counter counts clock pulses. The comparator matches the rising counter output (which has been converted to an analog voltage) against the analog input. When they become equal, the comparator output switches *low*, stopping the process. The counter holds that count until the SOC line goes *low* again.

The servo type of ADC shown in **Figure 10-18** hunts between two adjacent binary values. As analog input varies, digital output seeks to track it. This duplicates the behavior of a servomechanism: hence the designation *servo*.

This converter can follow small changes quite rapidly: for example, it can make one-count changes as fast as the clock frequency. However, a servo ADC requires a full series of counters in order to display a full-scale change of input.

PARALLEL ADC

A parallel ADC, like that shown in **Figure 10-19**, is also called a *flash converter*. It needs a comparator for each unit of resolution, plus a decoder to change the seven comparator outputs into a matching 3-bit digital code.

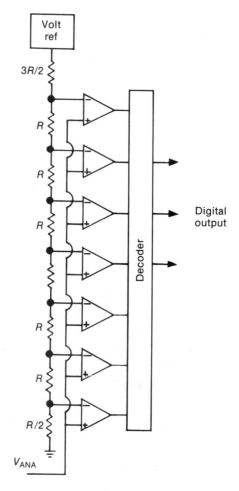

Figure 10-19. Parallel ADC or flash converter

A reference voltage connects to the negating inputs of the comparators. With 0 V analog input, all comparators send logic *lows* to the decoder, which provides a 000 digital output. As the analog voltage increases, each successive comparator switches on, so that at the upper voltage limit all the comparators are on. With each change in combined comparator output, the decoder provides another unique 3-bit output.

10
Digital and Analog Conversion

The advantage of this approach lies in its speed of conversion; only the switching times of the comparators and gates limit speed. Parallel A/D conversion is, in fact, the fastest approach.

One disadvantage is obvious: the number of comparators needed increases geometrically with each increase in resolution. So, a 3-bit converter takes seven op amps and anywhere from eight to ten decoder gates.

A converter less complex than the parallel ADC, and almost as fast, combines the parallel and SA techniques. It uses parallel conversion for a small number of bits, and SA for several bits. Referred to as modified parallel, this converter provides better resolution than plain parallel, and greater speed than SA.

SAMPLE-AND-HOLD METHODS

Modern ADCs are both reliable and accurate. However, maximum performance demands that analog-input voltage remain relatively constant in value and free of noise during the conversion period. Signal conditioning helps the signal meet the noise-free criterion; sample-and-hold circuits stabilize the signal for A/D conversion.

Sample-and-hold circuits, while not required for analog signals that vary slowly, become crucial to the accuracy of high-speed, high-resolution systems. The sample-and-hold concept is indicated by **Figure 10-20**. The circuit consists of a unity-gain input op amp, for isolation purposes, an analog switch (really an FET), a capacitor to hold the signal level; and another unity-gain output op amp to buffer the capacitor from the load.

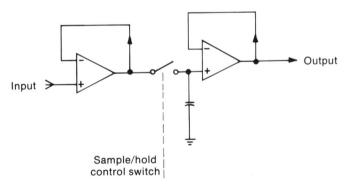

Figure 10-20. Sample and hold circuit

In its *sample* mode, the circuit passes the input signal directly to the capacitor, which tracks the analog voltage level. In the *hold* mode, the switch opens and the capacitor holds the latest value of the input signal. The load on the capacitor is

Digital Circuits and Devices

limited by the very high input impedance of an op amp. The op amp passes to the output whatever value the signal had reached at the time the switch opened.

Like DACs and ADCs, sample-and-hold circuits are readily available in standard packages.

Topic Review 10-5 and 10-6

1. The _____ type of A/D conversion is too slow for fast acquisition, but is used often in digital voltmeters.

2. The _____ type, double-ramp version cancels inherent errors.

3. The _____ type ADC is noted for both high resolution and high speed.

4. A single-ramp converter integrates a reference voltage that is opposite in polarity to the _____ signal.

5. Sample and hold circuits _____ the signal for A/D conversion.

Answers:

1. dual-slope
2. integration
3. SA
4. analog
5. stabilize

10-7 Summary Points

1. A DAC is a device that converts a binary count into a proportional analog output voltage.

2. The two types of DACs are the WRN and R2R ladder.

10
Digital and Analog Conversion

3. A monotonicity test is a visual inspection of a DAC's output to determine any malfunction in the circuit. A monotonicity test is dynamic, which means the test is performed under actual operating conditions.

4. The parallel or flash ADC is the fastest type of ADC.

5. An ADC is a device that converts an analog voltage into a binary count. The ADC allows external analog devices to communicate with a digital computer.

6. A ramp-type ADC uses an RC time constant to determine the output binary count.

7. A V/F ADC is used to count the number of pulses, which is used to indicate the analog input voltage.

8. A Servo ADC contains a counter, R2R, and comparator that is used to convert an analog input into a digital count.

9. A parallel ADC is very fast but requires too many parts.

10. Sample-and-hold circuits allow analog to digital conversion of inputs voltages that contain noise.

10-8 Chapter Progress Evaluation

1. What will be the resolution in a 5-bit R2R DAC, when $R_f = 10\ k$ and $R = 10$ kohms? A logic *high* equals 5 V.

2. What will be the resolution in a 4-bit R2R DAC, when $R_f = 10\ k$ and $R = 20$ kohms? A logic *high* equals 5 V.

3. A DAC that has resolution of 0.1 *V*/count has a higher resolution than a DAC with 0.5 *V*/count. True or False?

4. As resolution increases in a DAC, the output voltage per binary count will do what?

Digital Circuits and Devices

5. A DAC allows a computer to control external analog devices. True or false?

6. A monotonicity test is what type of test?

7. An ADC allows analog devices to talk to a digital computer. True or false?

8. What section in a servo-type ADC, produces the analog reference voltage?

9. The figure below shows the output of a 4-bit DAC, with a resolution of .1 V/count. Determine which resistor/s is/are missing when 2 = R1, 2 = R2, 2 = R3 and 2 = R4.

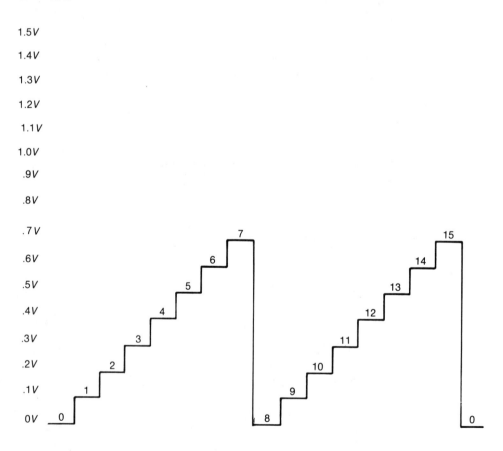

Chapter 11

Memory

Objectives

Upon completion of this chapter, you should be able to do the following:

* Describe the operation of various types of ROMs

* Describe the various types of static RAMs

* Describe the operation of dynamic RAM

* Describe the process of expanding the word size of a memory

* Describe briefly the operation of CCD, Josephson Junction, MBM, tape and floppy disk memory

* List the advantages and disadvantages of the various types of memories

11-1 Introduction

Before we begin learning about memory devices, this topic defines many new terms where most are related to computer systems. We will begin by covering the majority of the terms now, so you will not be confused later when we are talking about memory devices.

Access time: The amount of time required to perform a complete read operation. This is a measurement of a memory device's operating speed. It is the actual time between the memory receiving a read command and the data becoming available at the memory's output, ready for processing.

Address: A numerical value which designates a specific location in a memory device or the destination of the data in a computer system.

Byte: A group of bits processed together in parallel. Normally no matter what size the computer is, a byte is considered to be eight-bits.

Digital Circuits and Devices

Capacity: The total quantity of data that memory device can store, usually expressed in bits or bytes.

Abv.	Actual numbers of locations.
1K	$1024 = 2^{10}$
2K	$2048 = 2^{11}$
4K	$4096 = 2^{12}$
8K	$8192 = 2^{13}$
16K	$16,384 = 2^{14}$
32K	$32,768 = 2^{15}$
64K	$65,536 = 2^{16}$

Example: To find the total number of bits in a memory chip, multiply the word size times the number of memory locations.
A 2114 memory chip contains 1K × 4.
1024 × 4 = 4096 bits or memory cells.

$\overline{CS}$ (Chip Select): A control line which enables the output circuitry and allows the data to be transferred into and out of the memory or any other computer device. When $\overline{CS}$ is *low* the outputs will be enabled. This means data will be written into or read from the memory. When $\overline{CS}$ is *high* the output circuitry will be in a high impedance state, making it transparent to the computer system data bus.

Cycle time: (also known as read/write time) The length of time required to read data from a memory location, and then write new data back into the memory location.

Dynamic memory: A memory device where stored data must be rewritten into each memory cell periodically, even with power applied, otherwise data will be lost. Rewriting data into a memory cell is called the refresh operation.

EPROM: (Erasable Programmable Read Only Memory) A type of memory that is programmed by the user and can be erased and reprogrammed as often as desired. This memory is erased by the use of ultraviolet light applied through a window on the chip. Note all the memory locations will be erased at one time leaving a logic one in all memory locations. This type of memory is non-volatile which means the data is retained even with the removal of power.

Memory cell: An electronic circuit that is used to store a single bit of binary data (0 or 1).

Memory word: The number of bits in a sequence handled as one unit. This group of bits can be stored in one memory location. The larger the word the more precise and the more intricate the instruction.

Nibble: A group of four bits of data processed together. A nibble is half a byte, and it is the smallest group of bits a computer will handle as one unit. Nibbles are usually used when working with math problems.

Non-volatile memory: A type of memory device that does not require power to retain its data. If power is removed, the data is not lost. Examples are ROM, PROM and EPROM.

Programmable: The process of being able to change data or functions of a device automatically.

PROM: (Programmable Read Only Memory) A ROM that can be programmed only once after manufacturing. The programming of the PROM is done by the user by burning out a fuse in series with each memory cell. The process will create a logic one in the selected memory location. When the fuse is not burned open, the data is logic zero.

RAM: (Random Access Memory) A memory device that can be written into or read from by addressing any memory location when its control lines are at their proper levels. In this type of memory, the location of the data has no effect on access time of the data.

Random access: The process of obtaining data from computer storage with the access time independent of the location of the data.

Read operation: (Also called the FETCH OPERATION) The operation that senses the data in a memory location designated by an address, and then transfers a copy of the data to another location. Note in this type of operation, data is not changed or lost but remains in memory.

Refresh: An operation performed in dynamic memory. In this case, the data must be rewritten into each memory cell, so it is not lost due to leakage in the memory circuit.

RMM: (Read Mostly Memory) A type of data storage device like ROM, except it can be reprogrammed. The reprogramming of this type of memory is not normally changed constantly during normal operation like RAM. It is used like ROM which can be programmed for special purposes, and it is usually read from.

ROM: (Read Only Memory) A storage device where information is stored permanently. The data is stored or programmed only once, and it's performed at the factory during the manufacturing process. Note this type of memory is non-volatile. Also data can never be written into this memory by the user, just read from.

SAM: (Sequential Access Memory) A type of memory where access time is not constant, but varies depending on the address location of the data. In order to find the data required, all memory locations must be addressed sequentially until the proper location is found. This is a serial type of memory access which makes it slow as compared to random access memory.

Static memory: A memory device where stored data will remain permanently stored as long as power is applied. This is done without the need for periodically rewriting the data into the memory cell or refreshing the cell.

Volatile memory: A type of memory device that requires power to retain its data. If power is removed, all the data is lost.

Digital Circuits and Devices

$\overline{\text{WE}}$: (Write Enable) A control line which tells the memory chip the data on the data lines will be written into a memory location designated by the address line. When $\overline{\text{WE}}$ is low data will be written into memory, the data lines will act as inputs. When $\overline{\text{WE}}$ is *high*, data will be read from memory, the data lines will act as outputs.

Write operation: (Also called the STORE OPERATION) An operation whereby a new word is placed or stored in a memory location designated by the address lines. This new word is written into a memory location and replaces the word previously stored there.

11-2 Read Only Memory (ROM)

Read Only Memory (ROM) is the type of memory device where data can only be read from. The data is stored inside the device permanently when made by the manufacturer. Since the data can never be changed, it is known as a *non-volatile memory*. A non-volatile memory does not require power to retain its data. All ROMs allow data access to any memory location by placing an address on the address lines where access time is independent of the memory location in the device. This type of function is called *random access*. Therefore, a ROM refers to a random-accessed read only memory.

When a ROM is purchased, the data (program) is built into the integrated circuit. When a ROM is made, the manufacturer will place data into each memory location by either allowing or not allowing the memory cell device (some type of semiconductor) to be connected between a matrix row and column address location. The most popular semiconductor device today is normally a MOS FET which is used for slow, low power, and high density ROMs. Also available are bipolar transistor ROMs. These are used where high speed is the most important factor, and low density and high power dissipating is not as important. *Density* refers to the amount of semiconductor devices that can be placed on the integrated circuit in a specified area. In the early days of computers, diodes were used as the semiconductor device, but due to the large power dissipation, it is very rare to find diode ROMs today.

If the semiconductor device is connected between the selected row and column inside the matrix system, the data will be *low* (voltage dropped across the device forward biases the semiconductor) at the selected memory location. When the semiconductor is not connected between the selected row and column inside the ROM, the data will be *high* at the selected memory location.

When the ROM is made, the manufacturer will use a truth table supplied by the user in either paper, tape or disk form to determine the data to be stored in the ROM. In order for the manufacturer to change the data in the ROM, they must modify their photo-graphic negative (mask). This controls how the interconnections are made inside the chip during the manufacturing process of the integrated circuit. The mask

11
Memory

must be changed every time a new program is needed for a ROM. The process of changing the mask or program for the ROM is very time consuming for the manufacturer. As a result, this makes the cost to the user very expensive in small quantities (100 or less). The average cost for the first 100 ROMs depends on the type of ROM and the data inside. However, after the intital cost to change the mask, the price of the ROMs drops significantly depending on the quantity and type of ROM. Therefore, mask ROMs are used when large quantities of proven programs in ROMs are needed.

When testing (burning-in) a program that will be eventually placed inside a ROM, a different form of ROM is used. PROMs, EPROMs and EEPROMs are the types of ROMs which allow the user to program the ROM function in the field. PROMs can only be programmed once by the user, and they are used in the second phase of the burn-in process. The second phase is when the program has been tested in a lab environment and preliminary testing is required in an actual working environment. After a certain length of testing time, if no problems occur the data from the PROMs are used to make the masked ROMs. If problems do occur, then new PROMs are programmed with the new program changes. The reason why PROMs are used in the second burn-in phase is that the cost of the PROMs is about one tenth the cost of the other two types of programmable ROMs.

The initial phase of the burn-in process is when the program is first being developed. In this first phase of the burn-in process the program is entered into either EPROMs (Erasable Programmable ROMs) or EEPROMs (Electrical Eraseable Programmable ROMs). These two types of PROMs allow the user to program the ROM function and also allows the user to change the program as often as desired. This allows corrections to be made within the program if problems (bugs) occur. The EPROM uses ultra-violet light to erase all the data in the EPROM at one time. EPROMs cost less than EEPROMS, but more than PROMs and ROMs. EEPROMs (also called EAPROMs) allow the user to erase only one byte or word of data from the memory chip. The price of both EPROMs and EEPROMs limits these devices to development of the ROM programs or when only a few ROM type devices are needed. PROMs, EPROMs and EEPROMs will be covered in greater detail in section 11-3 RMM (Read Mostly Memory).

DIODE ROMS

Even though *diode ROMs* are very rare today, the basic operational concepts are used in all types of ROMs. The diagram of **Figure 11-1** shows two ROM memory cells where the semiconductor devices are diodes.

In order to select a row, it must go *low*. The data is then read from the column. In this example, column 2^0 will be *low*, because D_1 is connected to the row which is grounded. The output voltage from column 2^0 will be the voltage dropped across the

Digital Circuits and Devices

forward biased diode D_1. At the same time, D_2 is not connected to the row, and as a result it is not conducting. The output voltage at column 2^1 will be the supply voltage minus the voltage dropped across R_2.

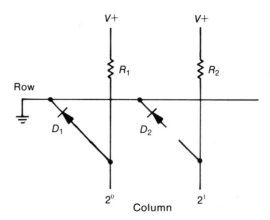

Figure 11-1. Two memory cells of a diode ROM

So each time there is a diode between a row and column and the row and column is selected, the data in the cell will be *low*. When there is no connection between the row and column, the memory cell will contain a logic *high*.

Figure 11-2 shows an 8 × 2 diode matrix ROM where 8 memory locations can be addressed each holding a word size of 2 bits. When there is no connection of the diode between the row and column, the open diode is not included in the drawing. Remember when the manufacturer makes the ROM, there is a diode in each cell. Some diodes are not connected to the rows allowing a logic one to be stored in that memory location.

In the diode matrix (8 ×2) ROM, address lines (A^0, A^1 and A^2) will select which row and column the data will be read from. When A^0 is *low*, columns A1 and A2 will apply their levels to the input of tri-state buffers 1 and 2. When A^0 is *high* columns B1 and B2 will apply their levels to the input of tri-state buffers 1 and 2.

The demultiplexer is used to select the row. It uses address lines A^1 and A^2 to make the selection. The demultiplexer circuit will have a *high* on all of its outputs except the select output which will be *low*.

When $A^1 = 0$ and $A^2 = 0$, row zero (R0) will go *low* and all other rows will be *high*. When $A^1 = 1$ and $A^2 = 0$, row one (R1) will go *low* while all other rows are *high*. When $A^1 = 0$ and $A^2 = 1$, row two (R2) will go *low* while all other rows are *high*. Finally when $A^1 = 1$ and $A^2 = 1$, row three (R3) will go *low*, while all other rows are *high*. In the rows that are *high*, when a diode is connected between the row and column the diode will be reversed biased. This means the column will sense an

11
Memory

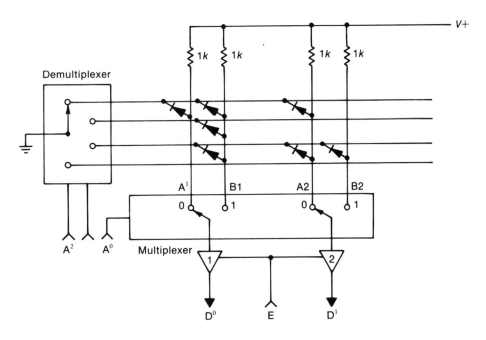

Figure 11-2. 8 × 2 diode matrix ROM

open circuit which does not effect the column data since all rows are in parallel with each other. When the row is *low* and a diode is connected between the row and column, the output of the column will be *low* (voltage dropped across the forward biased diode). The column will see the lowest voltage because all rows are in parallel with each other. Where there is no connection of the diode between the row and column, the column will see approximately the supply and the resistor will limit the current in the column. The data seen at the column will be a logic *high* (logic 1).

The last section of the diode matrix ROM is the tri-state buffers. They are used to allow more than one ROM to share the same data line. The data lines transfer data to and from the computer or other computer system devices. These data lines are called the *data bus*. All data inputs and outputs are connected to the same bus. In order not to connect more than one output to the same line at the same time, only one device is allowed to be enabled at any one time. All other devices will be disabled or in their high impedance state. When the output of a computer device is in its high impedance state, the output of a device looks like an open circuit. It is then transparent to any other device on the same line. The pin input that controls the tri-state output of each device is called E (Enable, active *high* because there is not negation sign above the letter) or $\overline{CS}$ (Chip Select, active *low* because there is a negation sign above the two

299

Digital Circuits and Devices

letters). For an active *low* control input (Enable or Chip Select), the output of the device will operate normally when the input is low. When the input is *high*, the output will be in a high inpedance state. For an active *high* control input (Enable or Chip Select), the output of the device will operate normally when this input is *high*. When the input is *low*, the output will be in a high inpedance state. In the diode matrix ROM diagram, the tri-state buffers will be enabled when the enable line is *high*. The outputs will be in the high inpedance state when the enable input is *low*.

Operation of Any ROM

The following sequence holds true for all types of ROMs.

1. Apply a new address to the address bus.
2. Enable the data outputs.
3. Read data from data outputs.
4. Disable data outputs before repeating the sequence.

Internal Operation of the Diode Matrix ROM

(Address $A^0 = 0, A^1 = 0, A^2 = 0$):

Row (R0) will be *low*, and column A1 will apply its voltage level to the input of buffer 1. Column A 2 will apply its voltage level to the input of buffer 2. Since there is a diode connection between row (R0) and column A1, column A1 will be *low*. There is also a diode connection between row (R0) and column A2, column A2 then will be *low*. When the enable line goes *high*, buffers 1 and 2 will output their input data(data = 00). When the data has been read, bring the enable line *low*. The output of the buffers will go into their high impedance state.

(Address $A^0 = 1, A^1 = 0, A^2 = 0$):

Row (R0) will be *low*, and column B1 will apply its voltage level to the input of buffer 1. Column B2 will apply its voltage level to the input of buffer 2. Since there is a diode connection between row (R0) and column B1, column B1 will be *low*. There is no diode connection between row (R0) and column B2; therefore, column B2 will be *high*. When the enable line goes *high* buffers 1 and 2 will output their input data (data = 10). When the data has been read, bring the enable line *low*. The output of the buffers will go into their high impedance state.

(Address $A^0 = 0, A^1 = 1, A^2 = 0$):

Row (R1) will be *low* and column A1 will apply its voltage level to the input of buffer 1. Column A2 will apply its voltage level to the input of buffer 2. Since there is no diode connection between row (R1) and column A1, column A1 will be *high*. There is also no diode connection between row (R1) and column A2; therefore, column A2 will be *high*. When the enable line goes *high*, buffers 1 and 2 will output their input data (data = 11). When the data has been read, bring the enable line low. The output of the buffers will go into their high impedance state.

11
Memory

(Address $A^0 = 1$, $A^1 = 1$, $A^2 = 0$):

Row (R1) will be *low*, and column B1 will apply its voltage level to the input of buffer 1. Column B2 will apply its voltage level to the input of buffer 2. Since there is a diode connection between row (R1) and column B1, column B1 will be *low*. There is no diode connection between row (R1) and column B2; therefore, column B2 will be *high*. When the enable line goes *high*, buffers 1 and 2 will output their input data (data = 10). When the data has been read, bring the enable line *low*. The output of the buffers will go into their high impedance state.

(Address $A^0 = 0$, $A^1 = 0$, $A^2 = 1$):

Row (R2) will be *low*, and column A1 will apply its voltage level to the input buffer 1. Column A2 will apply its voltage level to the input buffer 1, column A2 will apply its voltage level to the input of buffer 2. Since there is no diode connection between row (R2) and column A1, column A1 will be *high*. There is a diode connection between row (R2) and column A2; therefore, column A2 will be *low*. When the enable line goes *high*, buffers 1 and 2 will output their input data (data = 01). When the data has been read, bring the enable line *low*. The output of the buffers will go into their high impedance state.

(Address $A^0 = 1$, $A^1 = 0$, $A^2 = 1$):

Row (R2) will be *low*, and column B1 will apply its voltage level to the input of buffer 1. Column B2 will apply its voltage level to the input of buffer 2. Since there is a diode connection between row (R2) and column B1, column B1 will be *low*. There is a diode connection between row (R2) and column B2; therefore, column B2 will be *low*. When the enable line goes *high*, buffers 1 and 2 will output their input data (data = 00). When the data has been read, bring the enable line low. The output of the buffers will go into their high impedance state.

(Address $A^0 = 0$, $A^1 = 1$, $A^2 = 1$):

Row (R3) will be *low*, and column A1 will apply its voltage level to the input of buffer 1. Column A2 will apply its voltage level to the input of buffer 2. Since there is no diode connection between row (R3) and column A1, column A1 will be *high*. There is a diode connection between row (R3) and column A2; therefore, column A2 will be *low*. When the enable line goes *high*, buffers 1 and 2 will output their input data (data = 01). When the data has been read, bring the enable line *low*. The output of the buffers will go into their high impedance state.

(Address $A^0 = 1$, $A^1 = 1$, $A^2 = 1$):

Row (R3) will be *low*, and column B1 will apply its voltage level to the input of buffer 1. Column B2 will apply its voltage level to the input of buffer 2. Since there is no diode connection between row (R3) and column B1, column B1 will be *high*. There is also no diode connection between row (R3) and column B2; therefore, column B2 will be *high*. When the enable line goes *high*, buffers 1 and 2 will output their input data (data = 11). When the data has been read, bring the enable line *low*. The output of the buffers will go into their high impedance state.

Digital Circuits and Devices

Shown in **Table 11-1** is an operational truth table of this diode ROM. This table explains in short form the operation just given in the detailed internal operation of the diode matrix ROM. Note the data found in each memory location is identified by a specific address number.

Table 11-1.
Truth table of an 8 × 2 diode ROM

Address			Selected		Data		Enable
A^2	A^1	A^0	Row	Column	D^1	D^0	E
0	0	0	R0	A1 & A2	0	0	1
0	0	1	R0	B1 & B2	1	0	1
0	1	0	R1	A1 & A2	1	1	1
0	1	1	R1	B1 & B2	1	0	1
1	0	0	R2	A1 & A2	0	1	1
1	0	1	R2	B1 & B2	0	0	1
1	1	0	R3	A1 & A2	0	1	1
1	1	1	R3	B1 & B2	1	1	1

For each address selected in the diode ROM in this example, two bits of memory will be displayed on the data outputs. In a ROM with a word size of eight bits, each time an address is selected 8-bits of memory are displayed on the data outputs instead of just two bits. So even though the example ROM only displays two bits of data on its outputs, the concepts of operation are the same in larger ROMs.

BIPOLAR TRANSISTOR ROMS

As in the diode ROM, the *bipolar transistor ROM* selects each memory cell in the same fashion. The only difference between the two types of ROMs is that the semiconductor device for each memory cell is a bipolar transistor instead of a diode. The transistor inside each cell of a transistor ROM is set up to draw less current. Therefore a transistor ROM dissipates less power than a diode ROM. This is the main advantage for transistor ROMs.

Figure 11-3 shows two bipolar transistor memory cells, each of which are selected. Note the connections of the emitter, base and collector of each memory cell.

When the row is selected (brought *low*), column 2^0 will be *low* due to the forward biased transistor Q_1. The resistor connected in series with the emitter of Q_1 is to limit

11
Memory

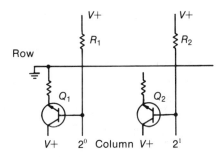

Figure 11-3. Two memory cells of a bipolar transistor ROM

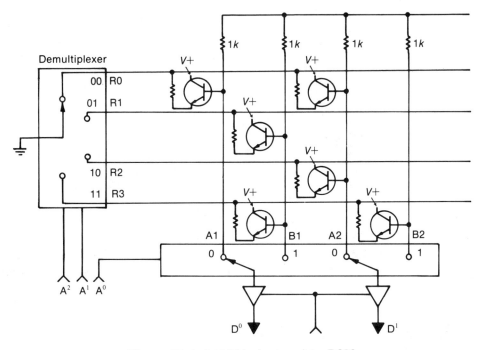

Figure 11-4. 8 × 2 bipolar transistor ROM

the transistor's emitter current. Note in transistor Q_2 the emitter circuit is not connected to the row. This transistor is not conducting in the circuit and will always be off. Column 2^1 will see a *high* because of the voltage applied to it from $V+$. As in the diode ROM, when a logic one is required for a memory location the manufacturer mask will change, so there is no connection between the emitter circuit and the row. **Figure 11-4** shows an 8 × 2 bipolar transistor ROM. To make the bipolar transistor

Digital Circuits and Devices

ROM easier to follow, when there is no connection between the emitter and row the transistor cell is left out of the drawing.

Shown in **Table 11-2** is the operational truth table for the bipolar transistor ROM. It should be noted that the rows and columns are selected in the same fashion as in the diode ROM.

Table 11-2.
Truth table of an 8 × 2 bipolar transistor ROM

Address			Selected		Data		Enable
A^2	A^1	A^0	Row	Column	D^1	D^0	E
0	0	0	R0	A1 & A2	0	0	1
0	0	1	R0	B1 & B2	1	1	1
0	1	0	R1	A1 & A2	1	1	1
0	1	1	R1	B1 & B2	1	0	1
1	0	0	R2	A1 & A2	0	1	1
1	0	1	R2	B1 & B2	1	1	1
1	1	0	R3	A1 & A2	1	1	1
1	1	1	R3	B1 & B2	0	0	1

The advantages of a transistor ROM over a diode ROM is that it is faster and dissipates less power. Also since the power dissipation is less, the manufacturer can place more memory cells in the same space. This increases the bit density of the chip. In general, bit density is inversely proportional to power dissipation.

MOS FET ROMS

As in the diode and transistor ROMs, the *MOS FET* (MOS for short) ROM selects each memory cell in the same fashion. The only difference is the type of semiconductor used to store each bit of memory. Shown in **Figure 11-5** is a PMOS memory cell. The substrate and source in this example are connected to the supply, through a current limiting resistor. The gate and drain are connected to each row. In this configuration, the PMOS FET is acting like an on or off switch. The PMOS FET is on when the row is brought *low*. This lowers the resistance of the FET to about zero ohms. The ground potential on the row is then applied to the column. When the

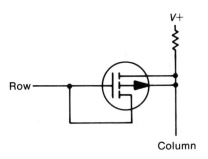

Figure 11-5. Basic PMOS memory cell

column is at ground potential, all the voltage is dropped across the column resistor. When the gate and drain circuits are *high*, as in an unselected row or in a cell that is not connected to the row, the resistance of the FET is more than 10 billion ohms. When the FET is off, the column will see the supply voltage from the column resistor and supply. If the cell that is selected is *low*, the column will see a *low*, because all rows are in parallel and the column is the common point. **Figure 11-6** shows two MOS memory cells that are both selected.

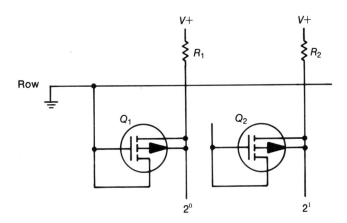

Figure 11-6. Two memory cells of a MOS FET transistor ROM

In this example, column 2^0 will be *low*, because a *low* is applied to the gate and

Digital Circuits and Devices

drain circuit of Q_1. This allows the resistance of Q_1 to be zero ohms and applies ground potential to the column. In column 2_1, there is no connection of the gate and drain circuit to the row. The transistor Q_2 is an open circuit, and a *high* will be applied to the column. To make the diagram shown in **Figure 11-7a** easier to analyze, when a cell is not connected to the row it will not be included in the diagram. This MOS FET ROM is an 8 × 2 where 8 memory locations can be addressed each holding a word size of 2 bits.

MOS ROMs have one main disadvantage as compared to diode and transistor ROMs. Because MOS FETs are voltage controlled devices and use a static charge to turn on and off, their operation is relatively slow (about 150 to 500 *nsec*). There are two advantages in using MOS ROMs. One is the very low power dissipation (about 1/100 the power needed for a transistor ROM). The other advantage is that MOS FETs take up less space; therfore, the bit density of the MOS ROM is very high on the order of 20 to 100 times more cells per area than transistor ROMs).

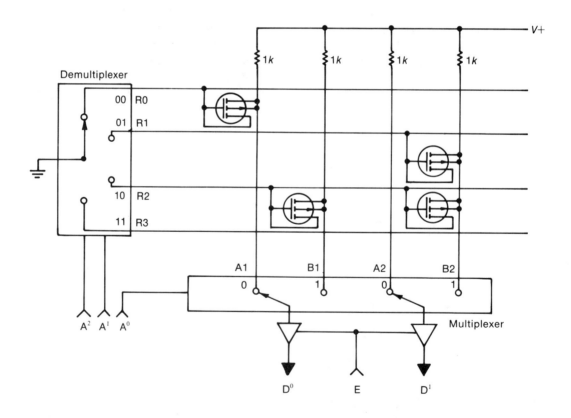

Figure 11-7. Diagram and truth table of an 8 × 2 MOS FET ROM

11 Memory

Address A^2 A^1 A^0			Selected Row	Column	Data D^1	D^0	Enable E
0	0	0	R0	A1 & A2	1	0	1
0	0	1	R0	B1 & B2	1	1	1
0	1	0	R1	A1 & A2	1	1	1
0	1	1	R1	B1 & B2	0	1	1
1	0	0	R2	A1 & A2	1	1	1
1	0	1	R2	B1 & B2	0	0	1
1	1	0	R3	A1 & A2	1	1	1
1	1	1	R3	B1 & B2	1	1	1

ROM ICS

Shown in **Figure 11-8** is the block diagram of an 8192 bit static MOS ROM. It is configured as a 1024 × 8-bit word memory chip.

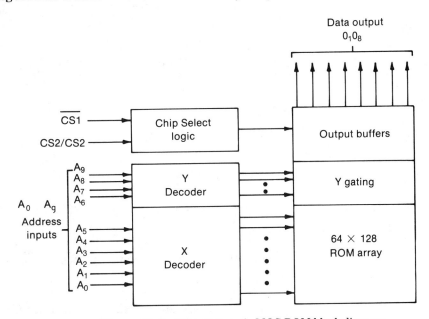

Figure 11-8. 1024 × 8 static MOS ROM block diagram

Digital Circuits and Devices

This ROM has ten address lines labeled A0-A9, and eight output data lines labeled O1-O8. Also included are two chip selects. $\overline{CS1}$ is fixed as an active *low* and CS2, the second chip select, is specified by the user as an active *high* or *low* at the time of manufacturing.

The timing diagram shown in **Figure 11-9** explains the time requirements to access this static MOS ROM chip.

Symbol	Parameter
t_{ACC}	Address to output delay time
t_{CO_1}	Chip Select 1 to output delay time
t_{CO_2}	Chip Select 2 to output delay time
t_{OF}	Chip Deselect to output data float time

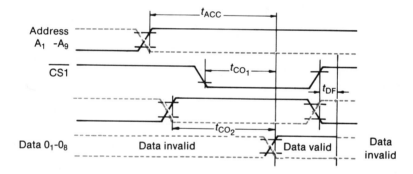

Figure 11-9. Timing diagram for a ROM

Topic Review   11-2

1. Data can only be read from a _____ .

2. Read Only Memory is a type of _____ memory since its data can never be changed.

3. The _____ _____ of a ROM is independent of the memory location in the device.

4. The type of ROM is identified by the _____ device making of the individual memory cells.

5. _____ _____ ROMs are the fastest whereas _____ _____ ROMs allow the greatest density.

Answers:

1. ROM
2. non-volatile
3. access time
4. semiconductor
5. Bipolar transistor, MOS FET

11-3 Read Mostly Memory (RMM)

Read Mostly Memory (RMM) devices retain data in a non-volatile state like ROMs, but they also have the ability to be programmed by the user. The programming of the RMM device is not a normal operation, and it requires special input voltage conditions.

PROMS:

The *Programmable Read Only Memory* (PROM) uses a row and column matrix memory cell system just like in standard ROMs. The main difference is a fuseable link in series with each memory cell semiconductor. This means all semiconductor devices are connected between the rows and columns in each memory location. Therefore, a new PROM has a logic zero in every memory location. The diagram of **Figure 11-10** shows two diode PROM memory cells. It should be noted the PROM cells may also be bipolar transistors or MOS FETs.

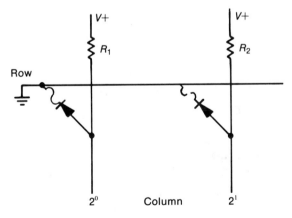

Figure 11-10. Two memory cells of a diode PROM

Digital Circuits and Devices

The memory cell connected between the row and column 2^0 will produce a logic zero on column 2^0. This is the condition of each new memory cell in a PROM before programming, or a cell that has been programmed for a logic zero. The memory cell connected between the row and column 2^1 will produce a logic one on column 2^1. The reason for the logic one at this column is that the fuseable link between the diode and column has been burned open in the programming process.

To program a PROM, a high voltage pulse of 15 to 30 volts with a time duration of 30 to $65 nsec$ is applied to the output data lines of the cells which a logic one is to be programmed into. The voltage will burn open the fuse of the memory cell selected. This will cause no conduction between the corresponding row and column. The column will then produce a logic one on its output when selected. Before any cell can be programmed, the memory location must be first addressed, and the chip selected with the chip select lines. If the memory cell is to have a logic zero in it, the voltage pulse is not applied to that particular data output line.

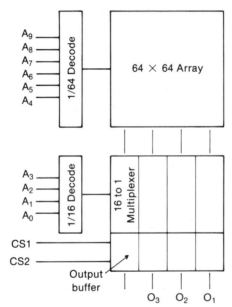

Figure 11-11. Bipolar transistor PROM block diagram

Once programmed, the PROMs program cannot be changed. If a mistake is made, the PROM cannot be used. For this reason, PROMs are normally programmed by computer controlled PROM burners. The *PROM burner* is simply an electronic device that allows a computer to automatically control the address lines, chip select, and data output lines. This is done in order to store the proper data in the proper memory locations of the PROM. Also if the voltage level or time duration of the pulse is not correct, the fuse may not open enough, and the logic one may change to a logic zero after a period of time. This process is called *reglow*, and it occurs after the chip has been programmed for a long period of time. The reason this occurs is that carbon forms in the burn area of the fuse. If the space of the open fuse is not large enough, carbon will build up during the operation of the PROM. Eventually the carbon will make a high resistance connection in the fuse resulting in the logic one to change into a logic zero. The newest types of PROMs use a polycrystalline silicon fuse which does not allow reglow as in the older types of PROMs.

Shown in **Figure 11-11** is the block diagram of a 1K × 4 bipolar transistor PROM. It has two chip select lines, ten address lines, and four output/program lines used for programming.

Figure 11-12 shows the timing diagram for this 1K × 4 bipolar transistor PROM.

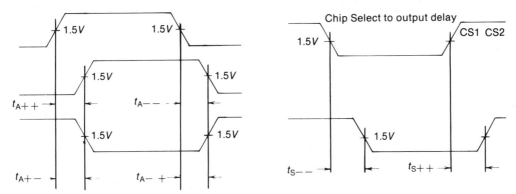

Figure 11-12. Timing diagram for a PROM

EPROMS

The *Erasable Programmable Read Only Memory* (EPROM) is a ROM type device that can be programmed, erased, and then reprogrammed as often as necessary.

The semiconductor memory cell device is a special FET with a light sensitive silicon gate circuit. It has a very high resistance when no light is applied to the chip or if it has not been programmed. The resistance of the silicon is so high that the FET

Digital Circuits and Devices

will not conduct even when selected. This means there is a logic one in every memory location of the EPROM before it is programmed.

To program data into the EPROM, the address must first be selected. Then the PGM (ProGraM enable logic pin) is brought to a TTL logic *high*. Next the data is placed on the data output (only TTL levels are needed). Then a pulse of proper level and time duration is placed on the VPP (Voltage Programming Pin). The VPP voltage which is between 15 and 40 volts will force electrons into the silicon gate region making the FET device conduct. Since there is negligible current flow in a FET gate circuit, the electrons will remain trapped in the gate region, and the FET will always conduct when selected. The process continues until all bits in the EPROM have been programmed. Before programming begins, a piece of non-transparent tape is placed across the erase window of the EPROM so that the data will not be erased by any light.

Once the EPROM is programmed, the only way to erase the data is to expose the window of the EPROM to ultraviolet light for a period of time. The light lowers the resistance of the silicon gate by producing photocurrent. This allows the trapped electrons to flow back into the silicon substrate. When enough electrons leave the gate region, the FET when selected will no longer conduct, and a logic one will be at the selected memory location.

The problem with EPROMs is that all memory locations are erased at the same time and the erase time can range from 1 hour to 6 hours. This time depending on the intensity of the light and the type of EPROM construction. EPROMs also have other disadvantages. They cost more than ROMs and PROMs. They're slower than PROMS and ROMs, and finally the erase window cannot be exposed to any type of light.

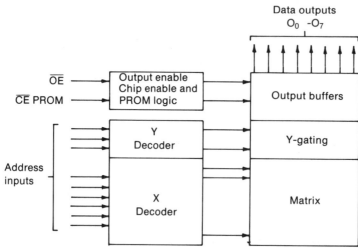

Figure 11-13. 2K × 8 EPROM block diagram

11
Memory

The main advantage of EPROMs is they can be reprogrammed as often as necessary and still function as an ROM. When used in circuits that only require a few ROM type devices, they are low in cost as compared to mask ROMs.

Shown in **Figure 11-13** is the block diagram of a 2K × 8 EPROM which can be programmed with TTL levels.

The timing diagram for this 2K × 8 EPROM is shown in **Figure 11-14**.

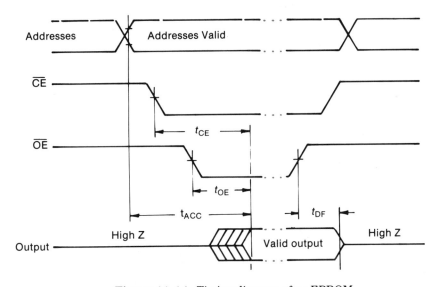

Symbol	Parameter
t_{ACC}	Address to Output Delay
t_{CE}	CE to Output Delay
t_{OE}	Output Enable to Output Delay
t_{DF}	Output Enable High to Output Float
t_{OH}	Address to Output Hold

Figure 11-14. Timing diagram of an EPROM

EEPROM

The *Electrically Erasable Programmable Read Only Memory* (EEPROM) is also called by some manufacturers EAPROM (Electrically Alterable Programmable Read Only Memory). The EEPROM is basically the same type of circuit as the EPROM.

Digital Circuits and Devices

However, instead of a light sensitive silicon gate, the EEPROM uses a silicon gate with an additional input. The input is an additional conductor. It is used to allow trapped electrons in the gate region to escape when erasing the memory location. In the EEPROM, only one word is erased at any one time instead of all memory cells as is usually the case with EPROMs. Just as in the EPROM, all cells have a logic one in them before programming.

To program a word into memory, the address and chip is first selected. Then a high voltage pulse of proper time duration is applied to the proper data output line. The voltage pulse forces electrons into the gate region which makes the FET conduct when selected. Since negligible current flows in the gate region, the electrons will remain permanently trapped.

To erase a word in memory, the adress and chip are again first selected. Then the proper voltage level is placed on the address erase pin. This voltage will allow the trapped electrons to leave the gate region. When enough electrons leave the region, the FET will no longer conduct when selected. A logic one will then result.

The advantage of the EEPROM over the EPROM is that one word can be erased instead of the entire memory chip. The disadvantages of EEPROMs are that they are slower than EPROMs and cost more.

Topic Review   11-3

1. An RMM stands for _____ _____ _____ .

2. An RMM is like a _____ in that is is non-volatile.

3. Unlike a ROM, the _____ has the ability to be programmed by the user.

4. Three basic types of RMM are _____, _____, and _____ .

5. The main difference between a ROM and a PROM is that with the _____ there is a fusable link in series with each memory cell semiconductor.

Answers:

1. Read Mostly Memory
2. ROM
3. RMM
4. PROMs, EPROMs, EEPROMs
5. PROM

11-4 Read Write Memory (RWM)

Read Write Memory (RWM) is the type of memory where data can be read from or written into. The data in a *static* RWM will be retained as long as the power is applied to the circuit. On the other hand, *dynamic* RWM stores its data in the capacitance of FETs, and as a result the data must be periodically re-established. Read Write Memory is usually called Random Access Memory (RAM). Therefore, a RAM refers to a random-accessed read write memory.

STATIC BIPOLAR TTL RAM

The *static bipolar TTL RAM* uses two crossed coupled TTL multi-emitter transistors that are configured as a latch to form each memory cell. The data will be retained inside the RAM as long as the power is supplied to the memory chip. If power is removed from the circuit, the data stored in the memory will be lost. This means that RAM is volatile.

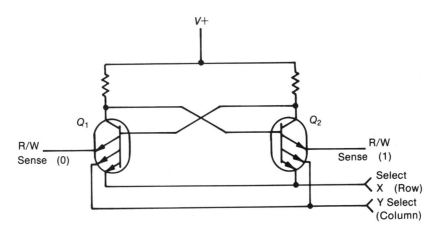

Figure 11-15. Static bipolar TTL RAM cell

Shown in **Figure 11-15** is a static bipolar TTL RAM cell.
In this memory cell, there are two crossed coupled multi-emitter transistors. In a multi-emitter transistor, any emitter can cause the transistor to conduct or keep it conducting. When power is first applied to this memory cell, one transistor will turn on first. This will apply a *low* to the base of the other transistor keeping that transistor turned off. The transistor which turns on first cannot be determined when power is first applied to the chip. As a result, the data in each cell is then invalid.

In order to select a static RAM memory cell, the row and column select lines must go *high*. This will force current to flow from either the sense amp or write amp circuit.

Digital Circuits and Devices

When the memory cell is deselected, the current required to keep the conducting transistor on will be taken from either the row or column select line. All row and column select lines will be *low* unless selected.

In the diagram of this basic static bipolar RAM cell when Q_1 is on, a *low* will be applied to the base of Q_2 keeping transistor Q_2 off. The data in the cell, when Q_1 is on and Q_2 is off, will be a logic zero. When Q_2 is on and Q_1 is off, the data in the memory cell will be a logic one. Since one transistor will always be on, this type of memory draws current even when data is not being read or written into the device. The resistor in series with the collectors limit the current through the conducting transistor.

The diagram of **Figure 11-16** shows a 4×1 TTL static RAM memory. This device contains 4, 1-bit words. It should be noted that most static RAM chips have 4 to 8 bit words.

When the power is first applied to the circuit, one transistor will start conducting before the other transistor. There is no way of determining which transistor in each cell will start conducting first. So when power is first applied to the circuit, the data inside the memory chip is invalid. For our discussion, we will make assumptions as to which transistor is conducting and which transistor is not conducting.

In this 4×1 static TTL RAM, there is both a row and a column decoder. Each decoder will output a *high* only on its selected output. All other outputs will be *low*. The selection of the row and column is determined by the logic levels on the address lines (A^0 and A^1).

The read sense amp is a op-amp that is connected as a comparator. If the voltage applied to the inverting input is *high* and the voltage applied to the non-inverting input is *low*, the output of the read sense amp will be a logic *low*. When a *low* is applied to the inverting input and a *high* is applied to the non-inverting input, the output of the read sense amp will be a logic *high*.

The output of the sense amp is an input for the $\overline{WE}$ (Write Enable) circuit. The write enable circuit contains a tri-state buffer which allows data to be supplied to the chip select circuit. Also included in the write enable circuit is a tri-state inverter. It is used as an input from the chip select circuit. When the $\overline{WE}$ line is *low*, data from the chip select circuit will be applied to this tri-state inverter. This will allow data to be written into the selected memory cell. At the same time, the tri-state buffer is in the high impedance state which does not allow the data from the sense amp to effect the selected memory cell. When the $\overline{WE}$ line is *high*, the tri-state buffer will allow the data from the sense amp to output its data to the chip select circuit. At the same time, the tri-state inverter of the write amp is in the high impedance state which will not allow the sense amp to effect the memory cell. So, in order to write into a memory cell, the $\overline{WE}$ line must go *low*. When reading the data from a memory cell, the $\overline{WE}$ line must go *high*.

The Chip Select ($\overline{CS}$) circuit acts as a bilateral switch. When $\overline{CS}$ is *low*, data can

11
Memory

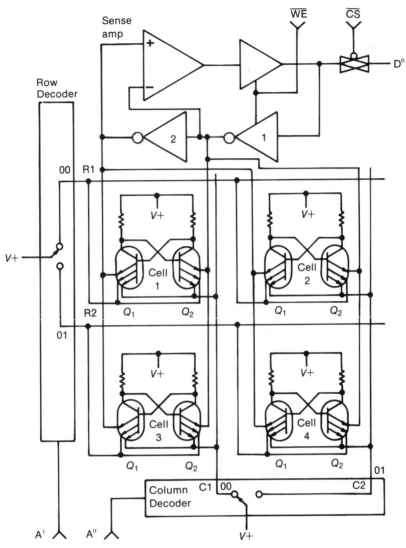

Figure 11-16. 4 × 1 static TTL RAM

flow into or out of the memory circuit. When $\overline{CS}$ is *high*, the output of the chip select circuit will be in a high impedance state. This allows more than one RAM output to share the same data bus as long as only one RAM chip is enabled at any given time.

To read data from a RAM, follow the steps listed below.

1. The $\overline{WE}$ line should stay *high*.

Digital Circuits and Devices

2. Apply the address to the address lines.
3. Apply a *low* to the $\overline{CS}$ line.
4. Read the data from the data bus.
5. Apply a *high* to the $\overline{CS}$ line.
6. Repeat steps 1 through 5.

To write data into a RAM chip, follow the steps listed below.

1. Apply the address to the address lines.
2. Apply a *low* to the $\overline{CS}$ and $\overline{WE}$ line.
3. Apply the data to be stored on the data bus.
4. Apply a *high* to the $\overline{CS}$ and $\overline{WE}$ lines.
5. Repeat steps 1 through 5.

Static TTL RAM Read Operation

We will assume transistor Q_2 of **Figure 11-16** will be conducting in memory cells 1 and 2. In memory cells 3 and 4, transistor Q_1 will be conducting. These conditions will occur when power is first applied to our example circuit.

Address $A^0 = 0, A^1 = 0, \overline{WE} = 1$ and $\overline{CS} = 0$:

When address line A^0 is *low*, column C1 will go *high*, and column C2 will go *low*. When address line A^1 goes *low*, row R1 will go *high*, and row R2 will go *low*. Since the current in memory cell 1 cannot get its current to keep transistor Q_2 conducting from the column or row, the current needed will be supplied by the sense amp. At this time, cell 2 is getting its current from column C2. Memory cell 3 is getting its current from row R2. Finally, cell 4 is getting its current from column C2 and row R2. Since Q_2 is conducting, a *low* is seen at the inverting input of the sense amp. The *low* at the collector of Q_2 is applied to the base of Q_1 which will keep Q_1 turned off. Since Q_1 is off, a *high* is applied to the non-inverting input of the sense amp. With the given input levels to the sense amp, the output of the sense amp will be a logic *high*. Since $\overline{WE}$ is *high*, the *high* from the sense amp is applied to the chip select circuit. The $\overline{CS}$ line is *low*, so the *high* is applied to the data line (D^0 = logic 1). Once the data has been read, the $\overline{CS}$ line will go *high*. This will cause the data bus to go into a high impedance state. When the $\overline{CS}$ circuitry is in the high impedance state, the output of the circuit becomes transparent to the data bus.

Address $A^0 = 1, A^1 = 0, \overline{WE} = 1$ and $\overline{CS} = 0$:

When address line A^0 is *high*, column C2 will go *high*, and column C1 will go *low*. When address line A^1 goes *low*, row R1 will go *high*, row R2 will go *low*. Since the current in memory cell 2 cannot get its current to keep transistor Q_2 conducting from the column or row, the current needed will be supplied by the sense amp or write amp.

Memory

At this time, cell 1 is getting its current from column C1. Memory cell 3 is getting its current from row R2 and column C1. Finally, cell 4 is getting its current from row R2. Since Q_2 is conducting, a *low* is seen at the inverting input of the sense amp. The *low* at the collector of Q_2 is applied to the base of Q_1. This will keep Q_1 turned off. Since Q_1 is off, a *high* is applied to the non-inverting input of the sense amp. With the given input levels to the sense amp, the output of the sense amp will be a logic *high*. Since $\overline{WE}$ is *high*, the *high* from the sense amp is applied to the chip select circuit. The $\overline{CS}$ line is *low*, so the *high* is applied to the data line (D^0 = logic 1). Once the data has been read, the $\overline{CS}$ line will go *high*. This will cause the data bus to go into a high impedance state making it transparent to the data bus.

Address $A^0 = 0$, $A^1 = 1$, $\overline{WE} = 1$ and $\overline{CS} = 0$:

When address line A^0 is *low* column C1 will go *high*, and column C2 will go low. When address line A^1 goes *high*, row R2 will go *high*, and row R1 will go *low*. Since the current in memory cell 3 cannot get its current to keep transistor Q_1 conducting from the column or row, the current needed will be supplied by the sense amp or write amp. At the same time, cell 1 is getting its current from row R1. Memory cell 2 is getting its current from row R1 and column C2. Finally, memory cell 4 is getting its current from column C2. Since Q_1 is conducting, a *low* is seen at the non-inverting input of the sense amp. The *low* at the collector of Q_1 is applied to the base of Q_2. This will keep Q_2 turned off. Since Q_2 is off, a *high* is applied to the inverting input of the sense amp. With the given input levels to the sense amp, the output of the sense amp will be a logic *low*. $\overline{WE}$ is *high*, and the *low* from the sense amp is applied to the chip select circuit. The $\overline{CS}$ line is *low*, so the *low* is applied to the data line (D^0 = logic 0). Once the data has been read, the $\overline{CS}$ line will go *high*. This will cause the chip select circuitry to go into a high impedance state which makes it transparent to the data bus.

(Address $A^0 = 1$, $A^1 = 1$, $\overline{WE} = 1$ and $\overline{CS} = 0$):

When address line A^0 is *high* column C2 will go *high*, and column C1 will go *low*. When address line A^1 goes *high*, row R2 will go *high*, and row R1 will go *low*. Since the current in memory cell 4 cannot get its current to keep transistor Q_1 conducting from the column or row, the current needed will be supplied by the sense amp of write amp. At this time, cell 1 is getting its current from row R1 and column C1. Cell 2 is getting its current from row R1. Finally, cell 3 is getting its current from column C1. Since Q_1 is conducting, a *low* is seen at the non-inverting input of the sense amp. The *low* at the collector of Q_1 is applied to the base of Q_2. This will keep Q_2 turned off. When Q_2 is off, a *high* is applied to the inverting input of the sense amp. With the given input levels to the sense amp, the output of the sense amp will be a logic *low*. Since $\overline{WE}$ is *high*, the *low* from the sense amp is applied to the chip select circuit. The $\overline{CS}$ line is *low*, so the *low* is applied to the data line (D^0 = logic 0). Once the data has been read, the $\overline{CS}$ line will go *high*. This will cause the data bus to go into a high impedance state making the chip transparent to the data bus.

Digital Circuits and Devices

Given the assumed values of the 4 x 1 TTL static RAM, the data stored in each of the four memory locations is tabulated in **Table 11-3**.

Table 11-3.
Truth table for the TTL static read operation example

Address		Data
A^1	A^0	D^0
0	0	1
0	1	1
1	0	0
1	1	0

Static TTL RAM Write Operation

Now that the data in the RAM has been read, let's demonstrate how to write into a memory location. From the previous explanation, we know the data in memory location 00 is a logic one. This means transistor Q_2 is on and Q_1 is off. In the following procedure, we will change the data in memory location 00 to a logic zero. This procedure is repeated any time you wish to change the data in a memory location. The only thing that changes is the address and the data.

Address $A^0 = 0$, $A^1 = 0$ and Data $= 0$:

Place the address on the address line, $A^0 = 0$ and $A^1 = 0$. When A^0 is *low*, column C1 will go *high*, and column C2 will go *low*. When A^1 is *low*, row R1 will go *high*, and row R2 will go *low*. Since memory cell 1 cannot get its current to keep transistor Q_2 conducting from the column or row, the current needed will be supplied by the sense amp. At this time, cell 2 is getting its current from column C2. Cell 3 is getting its current from row R2. Finally, cell 4 is getting its current from column C2 and row R2. Since Q_2 is conducting, a *low* is seen at the inverting input of the sense amp. The *low* at the collector of Q_2 is applied to the base of Q_1 which will keep Q_1 turned off. When Q_1 is off, a *high* is applied to the non-inverting input of the sense amp. With the given input levels of the sense amp, the output of the sense amp will be a logic *high*.

Next, the $\overline{CS}$ and $\overline{WE}$ lines are brought *low*. This will disable the tri-state buffer for the $\overline{WE}$ circuit. The data from the sense amp will not effect the chip select circuit, because the tri-state buffer is in its high impedance state. The new level is now placed on the data line (D^0). The new data is a logic zero. The *low* from D^0 is applied to the chip select circuit which is enabled at this time since $\overline{CS}$ is *low*. The *low* from the chip select circuit is applied to the tri-state inverter of the write enable circuit which is enabled. The *high* from the output of this tristate inverter is applied to the emitter of Q_2 in cell 1. At the same time, the *high* from the tri-state inverter is applied to inverter 2 which will output a logic *low* to the emitter of Q_1 in memory cell 1. With the emitter of Q_2 *high* and the emitter of Q_1 *low* in cell 1, transistor Q_2 will turn off and Q_1 will turn on. The transistor that was conducting in memory cell 1 has changed from Q_2 to Q_1. The data has also changed in memory cell 1 from a *high* to a *low*. Next the $\overline{WE}$ and $\overline{CS}$ lines are brought high again. This disables the RAM from the data line and allows the sense amp data to be applied to the input of the disabled chip select circuitry. The write operation can now be repeated again until the necessary memory locations have been written into.

The advantages of static TTL RAM are that it is fast and does not require any additional circuitry. The disadvantages of TTL static RAM are low bit density, high power dissipation, and the cost is high.

STATIC MOS RAM

The main difference between TTL and static MOS RAM is that static MOS RAM uses crossed coupled MOS FETs to latch the data into each cell.

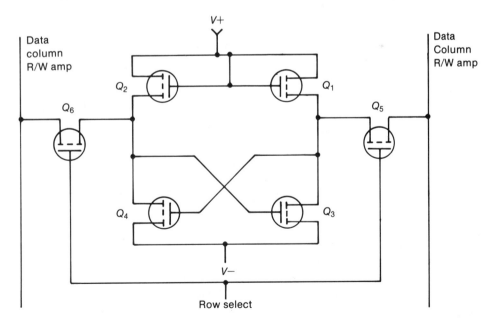

Figure 11-17. PMOS static MOS RAM cell

Digital Circuits and Devices

Shown in **Figure 11-17** is a PMOS static MOS RAM cell. Q_1 and Q_2 act as series limiting resistors to limit the current through the cell. Q_3 and Q_4 act as the crossed coupled FETs which form a latch circuit. When Q_3 is on, Q_4 will be off, and the data in the cell is a logic 1. When Q_4 is on, Q_3 will be off, and the data in the cell will be a logic 0. Q_5 and Q_6 are now switches that turn on when the row is selected. This allows the levels from Q_3 and Q_4 to appear on the column for each cell on the selected row.

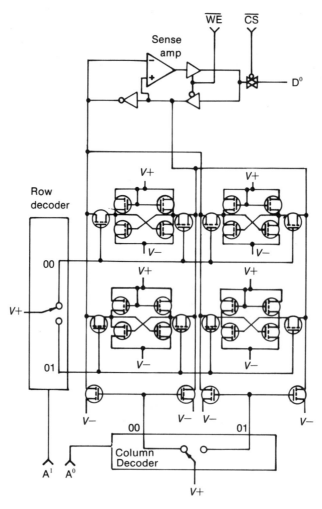

Figure 11-18. 4 × 1 NMOS static MOS RAM

Shown in **Figure 11-18** is a 4 × 1 NMOS static MOS RAM. Again each MOS RAM cell contains two row switches. When the gate circuit is brought *high* from the row decoder, the levels within each memory cell will appear on the columns for each cell on the selected row.

Since the data will appear on all columns of the circuit for the selected row at the same time, a MOS RAM uses column switches controlled by the column decoder to determine which columns' data will be connected to the sense amp or write circuitry. When the column decoder selects column C1, the gate to the column FETs will go *high*. This will allow the data from C1 to be applied to the sense amp or write circuitry. At this time, column C2 switches will have a *low* applied to the gate circuit. This will keep the two column FETs of C2 off. When the column FETs are off, the data on the unselected columns will not be affected.

The rest of the circuit works the same as in the TTL static RAM. Remember as in the TTL RAM, the data in the MOS RAM cannot be determined when power is first applied to the circuit.

Since MOS FETs dissipate less power and take up less space, the bit density of MOS RAM is greater than that of TTL RAM. The disadvantage of MOS RAM is that it is slower than TTL RAM.

Static RAMs can also be made with ECL or CMOS logic. The advantage of ECL memory is that it's the fastest type of memory. The disadvantages of ECL memory are that it dissipates more power than any other type of memory and that it has the lowest bit density. The advantages of CMOS memory are very high bit density and very low power dissipation. The disadvantage of CMOS is that it is relatively slow.

DYNAMIC RAM

Dynamic RAM stores its data in the capacitance of FETs. Because the capacitance in FETs is very small, the charge on the capacitance will dissipate very fast. Since the charge on a dynamic cell dissipates fast, the charge must be re-established often to maintain the data in each cell. The process of re-establishing the data in a dynamic memory cell is called the *refresh operation*. Dynamic RAM is made of MOS technology only because of the low power dissipation and current requirements.

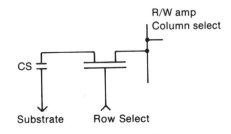

Figure 11-19. Dynamic MOS RAM cell

Shown in **Figure 11-19** is a diagram of a dynamic MOS RAM cell. Note it only requires one MOS FET which acts as an on and off switch. When the row select is

Digital Circuits and Devices

brought *high* the FET is on, and the charge level on the internal capacitance of the substrate will appear on the column. When the row select is *low*, the FET is off, and no level will be applied to the column.

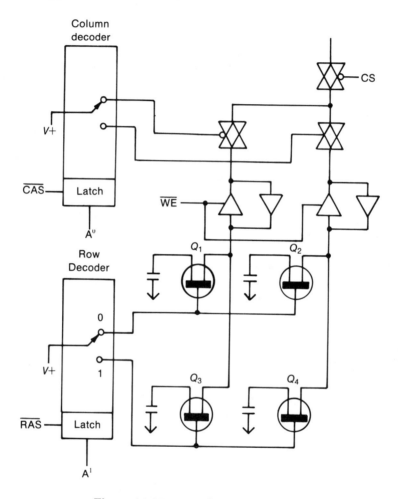

Figure 11-20. 4 × 1 dynamic MOS RAM

Shown in **Figure 11-20** is a diagram for a 4 x 1 dynamic MOS RAM. As can be seen from this dynamic RAM, each column has its own read/write circuit. One buffer or both buffers for each column will always be working.

To address a memory cell, the row address is placed on the address bus. Then the $\overline{RAS}$ (Row Address Strobe) is brought *low*. When the $\overline{RAS}$ goes *low*, the *high* to *low* transistion will cause the row address to be latched into the row decoder. The row

Memory

decoder will cause the selected row to go *high* while making all other rows *low*. The selected row will allow all memory cells in the row to apply its data in the form of a charge level to each column.

Next the computer places the column address on the address bus. Then the $\overline{CAS}$ (Column Address Strobe) is brought *low*. When the $\overline{CAS}$ line goes *low*, the *high* to *low* transition will cause the column address to be latched into the column decoder. The column decoder will select which one of the column read/write buffers will apply its data to the chip select circuitry.

The computer next selects whether the read buffer will be enabled or disabled. When $\overline{WE}$ is *low*, the read buffer will be disabled. It will allow the data from the chip select to be applied to the write buffer of the selected column. This process will cause the write buffer to establish the charge level on the memory cell that is selected. When $\overline{WE}$ is *high*, the read buffer for each column will be enabled. The data from each memory cell of the selected row is applied to each column. The level from each column is applied to each read buffer. The output of each read buffer is applied to the input of the chip select circuit and the input of the write buffer which is always working. The output of the write buffer then re-establishes the level on each memory cell of the selected row. Thus when a read operation is performed, each cell level is re-established on the selected row.

Finally, the Chip Select ($\overline{CS}$) circuitry controls whether the data bus will be connected to the RAM. When $\overline{CS}$ is *low*, the data can be read from or written into the memory chip. When $\overline{CS}$ is *high* the memory chip becomes transparent to the data bus. This function allows more than one memory device to share the same data bus as long as only one chip is enabled at any one time. If more than one device is enabled at the same time, the internal circuitry will burn up.

As previously stated, the data for each dynamic RAM cell is stored in the internal capacitance of the MOS substrate. Since the capacitance is very small, the charge level on the capacitance will dissipate fast every 2 to 10 *msec*. In order for the data not to be lost or changed, each cell must be refreshed every 2 to 10 *msec*. The length of time between refreshing each cell is determined by the type of construction of the cell and the quality of the material. In general, the more expensive the dynamic RAM the longer each cell can work without refreshing.

Each cell requires 0.2 to 0.5 *usec* to refresh. If the dynamic memory contains 65,536 cells, the amount of time it would take to address each cell to refresh would be 13.107 to 32.768 *msec*. This amount of time is too long. The cells by this time have changed their levels. Also since the memory cannot be used during the refresh cycle, the computer could never use the memory. The problem is how can all cells in dynamic RAM be refreshed and still allow time between refresh cycles for the computer to use the memory? The following outlines the answer to this problem.

Recall from earlier in this section that in the read operation of the dynamic RAM, each time a row is selected each cell connected to the row will be refreshed automatically. Since the computer cannot use the dynamic RAM during refresh in order to refresh a 65,536 × 1 dynamic RAM, each row will be addressed with $\overline{WE}$ and $\overline{CS}$ held *high*. When $\overline{WE}$ is *high*, the memory will be in the read mode. When $\overline{CS}$ is *high*, the data bus will not affect or be affected by the dynamic RAM. As in any type of random access memory device, the row and column matrix of a dynamic RAM is kept as equal as possible. This means the number of rows usually equals the number of columns. This simplifies the design of the memory. In a 65,536 × 1 (64K × 1) dynamic RAM, there are 256 rows and 256 columns per row. This gives a total of 65,536 memory cell address locations.

Refresh is therefore performed by a method called *RAS only refresh*. A circuit called the refresh controller, which today is a single chip MPU (Micro-Processor Unit), is programmed to address each row of each chip. Since the column data is never used, the controller does not address the column. Also the memory device is kept in the read state and the data I/O is always disabled ($\overline{CS}$ = *high*). If it takes between 0.2 to 0.5μsec to access any address, when only the row address is applied to the memory chip, each cell in the row will be refreshed at exactly the same time. Therefore, the time needed to perform a total refresh of a 64K × 1 dynamic RAM requires between 51 to 128$nsec$ (256 row addresses times 0.2 or 0.5μsec). This leaves some time for the computer to use the RAM between each refresh cycle. During the refresh cycle, the refresh controller will not allow the computer to use the memory being refreshed. After the refresh cycle is over, the refresh controller will allow the computer to use the memory.

The refresh controller can perform refresh in two different ways. One method is called a *distributed refresh mode* which spreads the refresh of each row over the entire refresh cycle time. The other method is called a *burst refresh mode*. This method disables the computer from using the memory and refreshes all the rows one after another until the entire RAM has been refreshed. There are no advantages to which method of refresh is used. It depends upon the type of refresh controller that is used in the memory circuit.

In the early days of dynamic memory, the refresh controller was built discretely. This required a large number of logic gates which made old dynamic memory not very useful. Today's refresh controllers are contained in the programming of one single chip MPU which makes dynamic RAM very useful in today's computers. In the late 70's, a type of dynamic RAM was developed with a built-in refresh controller for each chip. It was called a hidden refresh. This made the dynamic RAM operate like a static RAM. The reason this type of memory never became popular is that the bit density was very low, the price per bit was very high, and the power dissipation for each memory chip was high.

Dynamic RAMs have four advantages. They have the highest bit density, the simplest circuit design, the lowest cost per bit than any other type of RAM, and it is faster than static MOS RAM. The disadvantage of dynamic MOS RAM is that it needs a refresh controller.

Topic Review 11-4

1. RAM refers to a random-accessed _____ _____ memory.

2. The two basic types of RAM are _____ and _____.

3. RAM is _____ since the data stored in this memory will be lost when power is removed.

4. The advantages of _____ _____ RAM are its fast operation and its ability to not require any additional circuitry.

5. _____ _____ RAM has many advantages over _____ _____ RAM.

Answers:

1. read write
2. static, dynamic
3. volatile
4. static TTL
5. Dynamic MOS, static MOS

11-5 Expanding Word Size

In most large memory IC chips, the word size is either 1 or 4 bits in length. However, most computers use words of 8 or 16 bits in length. The example provided in **Figure 11-21** shows how to expand a memory's word size by paralleling memory chips to form blocks of memory. In this example of memory expansion, there are four 2114 which are 1K = 4 static MOS RAMs. The computer will need eight data bits to operate. Therefore, each chip will supply one half of the word, so two chips will have to be enabled at the same time. The enabling of the chips is controlled by one address line (A10) which is controlling an address decoder. The output of this address decoder will enable the $\overline{CS}$ line of each chip.

Digital Circuits and Devices

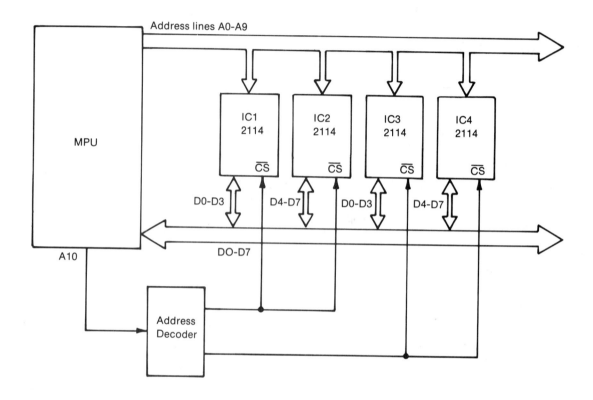

Figure 11-21. Example of memory expansion

When address line A10 is *low*, the address decoder will enable IC1 and IC2 which form memory block 1. Address lines A0-A9 will select which memory location in the block will provide its data to the data bus. IC1 will supply the lower nibble of the data word to data lines D0-D3. IC2 will provide the upper nibble of the data word to data lines D4-D7. Thus, the 8-bits required for the data bus is satisfied. At the same time, the address decoder will disable IC3 and IC4. This makes their data I/O lines transparent to the data bus.

When address line A10 is *high*, the address decoder will enable IC3 and IC4 while disabling IC1 and IC2. Memory block 2 will display its data on the data lines. The lower 4-bits are from IC3 and the upper 4-bits are from IC4. The memory location of the block will be determined by the address on address lines A0-A9. Therefore, we have expanded a 4-bit word to an 8-bit word memory.

In dynamic RAM each memory location usually contains a 1-bit word. Therefore,

11 Memory

to use dynamic RAM in an 8-bit computer, eight dynamic RAMs normally must be paralleled to obtain the 8-bit memory block.

11-6 Other Types of Solid State Memories

CHARGE COUPLED DEVICE (CCD)

The *Charge Coupled Device* (CCD) is a type of memory that uses MOS type shift registers in a write/recirculate mode. The CCD stores its data in the internal capacitance of the MOS registers which makes it act as a *dynamic memory*. Since CCD is a dynamic type of memory, refresh must be performed every 2 to 10 *msec*. Since the data is stored serially, refresh will occur when the CCD shifts one bit from one register to another. The CCD memory also is known as a Sequential Access Memory (SAM). This means, if you wish to access the data at memory location 100 in a CCD memory of 256 bits, it will take 156 clock pulses to get the data out of the device. If the data you wish to see is at memory location 200 in a CCD memory of 256 bits, it will take 56 clock pulses to get the data out of the device. As a result, the access time for this type of memory device will vary with the location of the data.

CCD comes in three forms. *Short loop* contains 64 to 512 bits of data. *Long loop* contains 1024 to 4096 bits of data, and *modified short loop* contains 4 to 8 short loops of 64 to 512 bits each. In this last type of CCD, there will be 4 to 8 data I/O lines.

CCD type memory is very slow as compared to random access type memory. *Low bit density* and the requirement of refresh to operate properly makes this type of memory not very popular in today's computer systems. This type of memory is used in some types of test equipment for short term data storage like in storage oscilloscopes.

JOSEPHSON JUNCTION MEMORIES

This type of memory configuration stores extremely large amounts of memory in a very small area. In fact, it can store 100 to 10,000 more data in the same area as the largest type of memory used in the computers today. The concepts behind the operation of *Josephson Junction* logic are to take low resistance materials like gold and lead and super cool them so there is no resistance in the circuit. Then by applying a magnetic field to a selected location, the current will begin to flow in the circuit. In theory, the current will flow forever with no resistance, and it will require no additional power to keep it there. To read the data, a magnetic pickup will detect which direction the current is flowing. When the current flows in one direction, the data is a logic 1. When the current flows in the opposite direction, the data is a logic 0. Also, since there is no resistance in the circuit to slow down the switching speed of the device, the switching time in the device is almost the speed of light.

Digital Circuits and Devices

The big drawback of a Josephson Junction type memory is that it must be cooled to a minus 271 degrees Celsius. This makes the operation very costly. Work is being done with Josephson Junction type memory to allow the circuit to work at higher temperatures with most of its advantages. When the temperature problem is solved, you will see a new type of computer system that will revolutionize the way computer systems operate. The advantages of Josephson Junction memories are that they operate almost at the speed of light and dissipate ten thousand times less power than the lowest power type of memory used today.

Topic Review   11-5 and 11-6

1. The _____ _____ of a memory can be expanded by paralleling memory chips to form blocks of memory.

2. The _____ _____ _____ or _____ is a type of memory that uses MOS shift registers in a write/recirculate mode.

3. A CCD memory is also known as a _____ _____ memory.

4. _____ _____ memories operate almost at the speed of light and dissipate extremely low amounts of power.

5. The _____ _____ type memory has the disadvantage of needing to be cooled to a minus 271 degrees Celsius.

Answers:

1. word size
2. charged coupled device, CCD
3. sequential access
4. Josephson Junction
5. Josephson Junction

11-7 Magnetic Memory Storage
MAGNETIC CORE MEMORY

This type of memory was one of the first types of magnetic memories to be used in computer systems. However, today *magnetic core memory* is not used because of the size of the units, the power requirements, the very complex interfacing circuitry, and the very very slow speed.

11
Memory

MAGNETIC BUBBLE MEMORY (MBM)

Magnetic Bubble Memory (Bubble Memory for short) stores its data in microscopic magnets and combines the read/write capability of RAM with the non-volatile function of ROM. Of the many companies that began working on MBM, only a few remain developing this type of memory. The main drawback is that MBM is very slow. It is almost as slow as the slowest tape system. In addition, the cost of the memory block and the control circuitry make it impractical as compared to other types of memory today. The major advantage of MBM is that it is a non-volatile read/write memory, and it requires very little power to retain its data.

The substrate or base of the MBM wafer is made up of a non-magnetic material called *gadolinium gallium garnet* (GGG). On one side of this substrate, a very thin layer of magnetic material is grown which is called the *epitaxial garnet*. Then on both sides of the MBM wafer, a strong permanent magnet is placed which sets up magnetic domains in the epitaxial garnet. When the domains of the epitaxial garnet oppose the magnetic fields of the permanent magnet, the shape of the garnet will change to form a bubble shaped object. This bubble represents a logic one. When the magnetic domains of the garnet do not oppose the magnetic fields of the permanent magnet, the shape of the garnet will be a cylinder shape which will represent a logic zero. The following diagram of **Figure 11-22** shows the reaction of the garnet under magnetic fields and the resulting logic states.

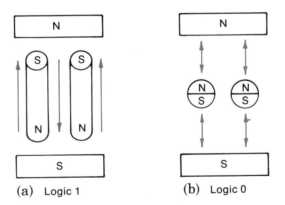

Figure 11-22. Logic state representation of the MBM wafer

The data in the wafer is created or changed by small electromagnetic fields. To create a bubble (logic 1) the magnetic field will turn the magnetic domains of the garnet so that the domain will oppose the flux of the permanent magnets. To change a bubble to a logic zero (no bubble), the magnetic fields of the write circuit will cause the magnetic domains of the garnet to turn. As a result, the domain does not oppose the flux lines of the permanent magnets.

Digital Circuits and Devices

Date is moved in the MBM wafer through the use of propagation tracks made of a soft ferro-magnetic material in the shape of *chevrons*. This type of magnetic material can be magnetized very easily; however, it will also lose its magnetism very easily. The domains of the magnetic garnet will line up under the propagation tracks.

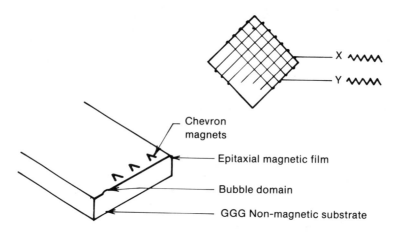

Figure 11-23. The MBM wafer

The part of the MBM wafer shown in **Figure 11-23** that controls the data movement is a pair of coils called the *orthogonal coil*. They are coils wound at right angles or perpendicular to each other. When a triangle wave is applied to each coil with a phase shift of 90 degrees, it will set up a magnetic field on the propagation tracks. This will cause the magnetic domains of the garnet to change their locations.

The data in the MBM is stored serially and flows in a loop configuration. As the magnetic domains of the garnet pass by the read detectors, the read detector will produce a signal. This signal will be converted into a logic one or zero, depending on whether the domains are opposing or aiding the magnetic flux of the permanent magnets.

The architecture of an MBM wafer is available in two forms: large loop and major/minor loop memory systems. In the large loop, one loop is used to store all the data. The loop may contain from 16K to 64K bits of memory. The large loop system was the first type of MBM wafer. It can require many clock pulses before the data is ready to be read. Also if any part of the loop is bad or goes bad, the entire loop will no longer function. **Figure 11-24** is an illustration of a large loop MBM system.

The system that is used today is called the major/minor loop system. The reason is it combines small loops which contain the data and one large loop which is used to transfer the data between the minor loops to the read and write detectors. One advantage of the major/minor loop system is that it combines random access of the minor loop and sequential access of the data in each minor loop. This process speeds up the access time of the MBM wafer. Also if one of the minor loops is defective, a

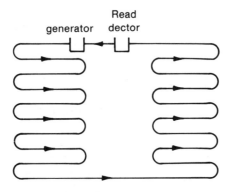

Figure 11-24. Large loop MBM system

PROM in the MBM controller will not allow data to be stored in the bad minor loop. Shown in **Figure 11-25** is an illustration of a major/minor loop MBM wafer.

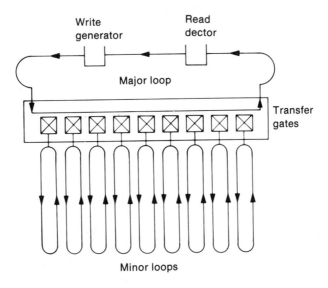

Figure 11-25. Major/minor loop MBM system

MBM type memories require both analog and digital control circuitry. The analog control circuitry drives the orthogonal coils and the read/write detectors all of

333

Digital Circuits and Devices

which require very precise timing. The digital circuitry is used to convert the analog levels of the read/write detectors. Digital circuitry also converts serial data into parallel data and parallel data into serial data to allow the computer to use the information stored inside the memory chip.

MBM memory is only used when very slow memory, very low power, and a non-volatile function are required.

MAGNETIC TAPE STORAGE

Magnetic tape recording memory is a memory storage system which stores data by recording and reading magnetic spots on a moving surface of magnetic material. The data is stored on the magnetic tape serially which makes tape storage a sequential type of access memory. In this type of system, only one bit of binary data is stored on the tape at any one time. There is also a modified type of sequential tape storage which stores one word of data (8 to 9-bits) at a time and is used in large computer systems. This modified SAM recording is also called multi-track recording, and it is 8 to 9 times faster than standard SAM recording.

The data may be recorded using analog signals. These signals use the intensity of the magnetic fields and the rate of change in the magnetic fields to determine the data stored. In digital recording, the rate of change and the direction of the flux changes determine the data stored on the tape. Digital recording is by far the more reliable type of tape storage, since analog recording is affected by noise and level changes.

The advantages of magnetic tape storage are very low cost, non-volatile, and can be reprogrammed many times. The disadvantage of tape storage is the very slow access time. Also on multi-track systems, the complexity of the mechanical parts increases the price of the equipment.

FLOPPY DISK STORAGE

A *floppy disk system* stores data on a disk shaped piece of thin mylar which is coated with a magnetic material. As the disk rotates, the data is stored sequentially in small groups of 128 to 512 bits which are called sectors. Each block of data is then random placed throughout the disk on tracks. The control circuitry along with a computer program called DOS (Disk Operating System) determines where the data will be stored and how to retrieve it. Because the sectors or data are random accessed, the speed of a floppy disk system has very fast access time. The amount of data that can be stored on a $5\frac{1}{4}$ inch disk can range from 80K bytes up to 500K bytes. On an 8 inch disk, the amount of data can range from 256K to 1.2M bytes.

The advantages of disk storage are very large storage in a small area, very fast, very little cost of the disk, and very low error rate. The main disadvantages of a disk

11
Memory

storage system are initial cost of the hardware and some mechanical breakdown tendencies.

Topic Review   11-7

1. _____ _____ memory stores data in microscopic magnets.

2. The main drawback of _____ is that it is very slow.

3. _____ _____ memory is a non-volatile read/write memory that requires very little power to retain its data.

4. The data in an MBM wafer is created or changed by small _____ _____ .

4. A _____ _____ system stores data on a disk shaped piece of thin mylar coated with a magnetic material.

Answers:

1. Magnetic bubble
2. MBM
3. Magnetic bubble
4. electromagnetic fields
5. floppy disk

11-8 Summary Points

1. Read Only Memory (ROM) is a random access type of memory that is non-volatile, and its data can only be read from. The different types of ROMs are identified by the semiconductor device making up the individual memory cells. The three basic types are diode, bipolar transistor, and MOS FET ROMs.

2. Diode ROMs exhibit high power dissipation, and they are rarely used today. Bipolar transistor ROMs exhibit high speed, but have lower density and higher power dissipation than MOS ROMs. Finally, MOS ROMs are mainly used for slow, low power dissipation, and high density ROMs.

3. Read Mostly Memory (RMM) is a random access type of memory that is non-volatile just like a ROM; however, a RMM also has the ability to be programmed by the user. The three basic types of RMMs are PROMs, EPROMs, and EEPROMs.

Digital Circuits and Devices

4. A PROM (Programmable Read Only Memory) is like a standard ROM except it contains a fuseable link in series with each semiconductor device within each memory cell. This type of RMM can be programmed only once by the user.

5. An EPROM (Erasable Programmable Read Only Memory) is a ROM type device that can be programmed, erased, and then reprogrammed as often as necessary. The semiconductor device is a special FET with an ultraviolet light sensitive silicon gate circuit. The data within an EPROM is erased by ultraviolet light.

6. An EEPROM (Electrically Erasable Programmable Read Only Memory) is very similar to an EPROM except it uses a silicon gate with an additional input instead of a light sensitive silicon gate. EEPROMs are slower than EPROMs.

7. Read Write Memory (RWM) is a random access type of memory where data can be read from or written into. A RWM is volatile, and it is usually called a Random Access Memory (RAM). The two types of RAM are static and dynamic.

8. The word size of a memory can be expanded by paralleling memory chips to form blocks of memory.

11-9 Chapter Progress Evaluation

1. What is the name of the process of obtaining data from a memory device where access time is independent of the memory location?

2. What is the word used to describe a memory device that doesn't require power to retain its data?

3. What type of memory acts like a ROM but can be reprogrammed?

4. Which type of ROM has the highest bit density?

5. Which type of ROM is the fastest?

6. Is static RAM a volatile memory?

7. What does a static MOS RAM cell act as?

8. What will happen to the data in a dynamic RAM cell when refresh does not occur?

9. What are the major advantages of MBM?

Chapter 12

Introduction to Computers

Objectives

Upon completion of this chapter, you should be able to do the following:

* Define basic computer terminology

* Describe the operation of the 74181 ALU as it relates to each function

* List the two methods of increasing word size in an ALU

* Describe the function of each block of the 6802 MPU IC.

* Briefly describe the function of each section of the block diagram of the D5 kit system
* Briefly describe the funtion of each section of the block diagram of the D5 kit system

* Describe the purpose of a memory map

12-1 Introduction

As an introduction to this chapter, let's define some of the basic terminology associated with computers.

ACC: (ACCumulator) It is a register that holds data used by the ALU for some type of functional change. The accumulator also holds the answer from the ALU after the function has been performed.

ACIA: (Asynchronous Communication Interface Adapter) It is a device that converts parallel data from the data lines into serial data which allows data transfer to an external serial device. The ACIA also performs the opposite function.

Address bus: They are control lines used to determine which device or memory

Digital Circuits and Devices

location will be accessed. These are unidirectional lines that act as inputs to all devices in the system except for the MPU, where they are used as outputs.

ALU: (Arithmetic Logic Unit) It is a device inside the MPU IC that contains the microprocessor data processing logic. The following are typical examples of functions that the ALU can perform.

A) Add
B) Subtract
C) AND
D) OR
E) ex-OR
F) Complement
G) Shift Right
H) Shift left
I) Increment
J) Decrement

Control logic: It is a circuit that generates the signals necessary to execute an instruction. This circuit is found inside the MPU IC.

Data bus: They are control lines used to transfer data from one device or memory location to another in a computer system. The data bus is bidirectional which means the data lines can be used as inputs or outputs.

Instruction cycle: It is a fetch (load) and execute operation for an MPU.

Instruction register: It is a register that holds the instruction being executed (OP code) by the MPU. This register is found inside the MPU IC.

Instruction decoder: It is a circuit inside the MPU that takes the instruction code from the instruction code register and then decodes it and applies the respective signals to the control logic. This circuit allows an instruction to be automatic.

Interrupt: It is a signal to the MPU that causes the MPU to halt current program execution, saves the internal environment (via hardware and/or software), and executes a program to service the signal.

Internal environment: It refers to the data in the InDex register, Program Counter, ACCumulator A, ACCumulator B, and Condition Code Register. These registers contain data that is being used by the MPU.

ID: (InDex register) It is a register that holds a quantity which may be used to modify an address or may be used for other purposes as set by the program.

MICRO: (MICROprocessor also called MPU) It is an integrated circuit capable of interpreting, and with the help of memory and input/output devices, it executes instructions automatically.

Memory address register: It is a register inside the MPU which will determine the address on the address lines until the device can be accessed. The address in this register is the one currently being read from or written into at the time.

Monitor program: (D5BUG in the D5 kit) It is a set of instructions in ROM which tells the MPU what to do. This program decodes input keys so a program can be loaded into memory. Also, it sets up everything in the MPU from resets, vectors, displays, and most other MPU controlled operations so the computer can run the program stored in memory.

PC: (Program Counter) It is a register that contains the address of the next byte of the instruction to be fetched from memory. The current value of the program counter is placed on the address bus, and then the program counter will increment automatically.

Programmed data transfer: Refers to the process where data is moved throughout the MPU. Data is first moved into the accumulators and then to its designation.

PIA: (Peripheral Interface Adapter) It is a device that allows the computer to communicate with the outside world in a parallel fashion.

CCR: (Condition Code Register) It is a register used to store certain results of a test performed during the execution of a program or ALU operation. The data in this register is used by the branch instructions to make decisions.

SP: (Stack Pointer) It is a register that contains the beginning address of the internal environment normally in RAM. Therefore, when an interrupt is received by the MPU, the data in the internal environment can be stored for later use. After the interrupt has been serviced, the MPU will use the data in the stack pointer to retrieve the stored data so that the program can continue from where it left off.

Subroutine: It refers to when the program leaves a sequential order of the main program to execute another program and then returns it to the intitial program.

Temporary data register: It is a register inside the MPU that stores data to be used in an ALU operation. This register is needed because the ALU has no memory storage capability.

12-2 Arithmetic Logic Unit

The Arithmetic Logic Unit (ALU) is the part inside a computer chip that performs mathematical and logic functions on the data the computer uses. In other words, the ALU is the calculating section of the computer system.

In our example, we will be discussing the 74181 4-bit MSI TTL ALU. The 74181 contains 32 separate operations that are divided into arithmetic and logic functions. The operations are performed on two 4-bit words labeled word A and word B and any level on the CN pin (ripple carry input). The result of function is a 4-bit word labeled F0-F3 (function outputs) and one additional output labeled CN+4 (ripple carry output). There is also another output called A=B. This output will go *high* when words A and B are equal without the selected function affecting the level of this pin.

Two or more ALUs may be placed in series with each other to increase the word size where the operation of each function will not change. The carry and borrow operations can be performed in two different ways. The simplest method is to use the ripple (serial) carry/borrow operation as illustrated in **Figure 12-1**. In this method the CN+4 (ripple carry) output of the least significant 4-bit word (nibble) or the least significant ALU is applied into the CN (ripple carry) input of the next ALU. Since this method uses a ripple carry/borrow bit, the operation of the function must be

Digital Circuits and Devices

complete before the bit can change its level. This makes the ripple carry/borrow function slow when two or more ALUs are connected together.

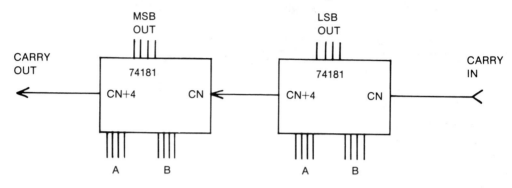

Figure 12-1. Expanding word size of an ALU using ripple carry

The second method of expanding the word size of an ALU uses the 74182 *look-ahead carry generator*. This device allows parallel generation of carry and borrow functions for up to four ALUs. By using the two additional outputs of the ALU, carry propagate (labeled P) and the carry generate (labeled G), parallel transfer of any carry or borrow from one ALU to another ALU is possible. This second method of expanding word size is shown in **Figure 12-2**. By using the 74182, the transfer rate of the carry/borrow bit can be increased by a factor of two to four times. This increases the speed of the operation of ALUs connected together.

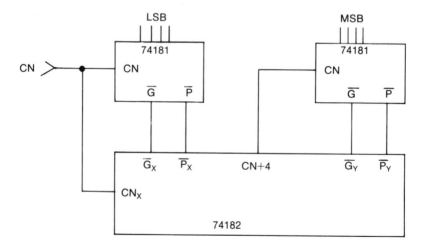

Figure 12-2. Expanding word size of an ALU using the 74182

Introduction to Computers

The ripple carry input (CN) and output (CN+4) act differently depending upon the type of operation that is being performed. When performing an operation in the logic mode, any carry in or out will not effect the ALU. When performing arithmetic operations, the ripple carry in (CN) and carry out (CN+4) will react differently depending on whether an addition or subtraction operation is performed. In performing addition, when CN+4 is *low*, a carry is represented; when CN+4 is *high*, a no carry condition is represented. In subtraction, when CN+4 is *low*, no borrow is represented; when CN+4 is *high*, a borrow is represented.

The operation that is performed is selected by the binary data on the four select lines (S0-S3) and the level on the mode control (M). The mode control (M) selects whether the sixteen logic or the sixteen arithmetic functions will be accessed by the four select line inputs.

Shown in **Table 12-1** is a listing of all the different functions of the 74181 ALU.

Table 12-1.
List of the functions of the 74181 ALU

Mode Select Inputs				Active High Inputs and Outputs	
				Logic	Arithmetic
S3	S2	S1	S0	(M = 1)	(M = 0) (CN = 1)
0	0	0	0	$\overline{A}$	A
0	0	0	1	$\overline{A + B}$	A + B
0	0	1	0	$\overline{A}B$	A + $\overline{B}$
0	0	1	1	Logic 0	minus 1
0	1	0	0	$\overline{AB}$	A plus $A\overline{B}$
0	1	0	1	$\overline{B}$	(A + B) plus $A\overline{B}$
0	1	1	0	A⊕B	A minus B minus 1
0	1	1	1	$A\overline{B}$	AB minus 1
1	0	0	0	$\overline{A}$ + B	A plus AB
1	0	0	1	$\overline{A \oplus B}$	A plus B
1	0	1	0	B	(A + $\overline{B}$) plus AB
1	0	1	1	AB	AB minus 1
1	1	0	0	Logic 1	A plus A
1	1	0	1	A + $\overline{B}$	(A + B) plus A
1	1	1	0	A + B	(A + $\overline{B}$) plus A
1	1	1	1	A	A minus 1

In the following examples of the 32 functions of the 74181 ALU, sixteen functions are in the *logic mode* and sixteen are in the *arithmetic mode*. When performing functions in the logic mode, CN is used to determine whether positive or negative

Digital Circuits and Devices

logic is used. When CN is *high*, we will be using positive logic. When CN is *low*, we will be using negative logic. In all of our logic mode examples, positive logic will be used.

LOGIC MODE

In the following logic mode functions, M=1, CN=1, word A=1001 and word B=0110. Remember that in this mode, the level on the CN+4 output is not a valid output for the ALU. It should also be noted that each bit in words A and B is treated individually when a function is performed.

EXAMPLE 1: (Select = 0000) (Function = $\overline{A}$)
The function states that the output of the ALU will be equal to word A with each bit inverted.
    A = 1001
    $\overline{A}$ = 0110
    output answer = 0110

EXAMPLE 2: (Select = 0001) (Function = $\overline{A + B}$)
The function states that the output of the ALU will be equal word A NORed to word B.
    A = 1001
    B = 0110
    output answer = 0000

EXAMPLE 3: (Select = 0010) (Function = $\overline{A}B$)
The function states that the output of the ALU will be equal word A inverted ANDed to word B.
    A = 1001
    $\overline{A}$ = 0110
    B = 0110
    output answer = 0110

EXAMPLE 4: (Select = 0011) (Functional = Logic 0)
The function states that the output of the ALU will be equal to a logic 0. The output is not affected by words A or B.
    output answer = 0000

EXAMPLE 5: (Select = 0100) (Functional = $\overline{AB}$)
The function states that the output of the ALU will be equal to word A NANDed to word B.
    A = 1001
    B = 0110
    output answer = 1111

EXAMPLE 6: (Select = 0101) (Function = $\overline{B}$)

The function states that the output of the ALU will be equal to word B with each bit inverted.
   B = 0110
   $\overline{B}$ = 1001
   output answer 1001

EXAMPLE 7: (Select = 0110) (Function A ⊕ B)
The function states that the output of the ALU will be equal to word A ex-ORed to word B.
   A = 1001
   B = 0110
   output answer 1111

EXAMPLE 8: (Select = 0111) (Function A$\overline{B}$)
The function states that the output of the ALU will be equal to word A ANDed to word B inverted.
   A = 1001
   B = 0110
   $\overline{B}$ = 1001
   output answer 1001

EXAMPLE 9: (Select = 1000) (Function $\overline{A}$ + B)
The function states that the output of the ALU will be equal to word A inverted ORed to word B.
   A = 1001
   $\overline{A}$ = 0110
   B = 0110
   output answer 0110

EXAMPLE 10: (Select = 1001) (Function $\overline{A \oplus B}$)
The function states that the output of the ALU will be equal to word A ex-NORed to word B.
   A = 1001
   B = 0110
   output answer 0000

EXAMPLE 11: (Select = 1010) (Function B)
The function states that the output of the ALU will be equal to word B.
   B = 0110
   output answer 0110

EXAMPLE 12: (Select = 1011) (Function AB)
The function states that the output of the ALU will be equal to word A ANDed to word B.
   A = 1001
   B = 0110
   output answer 0000

Digital Circuits and Devices

EXAMPLE 13: (Select = 1100) (Function Logic 1)
The function states that the output of the ALU will be equal to a logic 1 for each output bit. Words A and B will not effect the output of the ALU.
    output answer 1111

EXAMPLE 14: (Select = 1101) (Function A + $\overline{B}$)
The function states that the output of the ALU will be equal to word A ORed to word B inverted.
    A = 1001
    B = 0110
    $\overline{B}$ = 1001
    output answer 1001

EXAMPLE 15: (Select = 1110) (Function A + B)
The function states that the output of the ALU will be equal to word A ORed to word B.
    A = 1001
    B = 0110
    output answer 1111

EXAMPLE 16: (Select = 1111) (Function A)
The function states that the output of the ALU will be equal to word A.
    A = 1001
    output answer 1001

ARITHMETIC MODE

In the following arithmetic mode functions, M=0, CN=1, word A=1001 and word B=0110. It should be noted that each bit in each words A and B is treated individually when a function is performed. An arithmetic function is only performed when the function is spelled out. Any signs will be treated as logic functions. Remember the CN−4 output will change levels with the function of the ALU in this mode, and, as a result, this output will be valid.

EXAMPLE 1: (Select = 0000) (Function A)
The function states that the output of the ALU will be equal to word A.
    A = 1001
    output answer 1001
    CN+4 = 1 (no carry)

EXAMPLE 2: (Select = 0001) (Function A + B)
The function states that the output of the ALU will be equal to word A ORed to word B.
    A = 1001
    B = 0110
    output answer 1111
    CN+4 = 1 (no carry)

EXAMPLE 3: (Select = 0010) (Function A + $\overline{B}$)
The function states that the output of the ALU will be equal to word A ORed to word B inverted.
    A = 1001
    B = 0110
    $\overline{B}$ = 1001
    output answer 1001
    CN+4 = 1 (no carry)

EXAMPLE 4: (Select = 0011) (Function Minus 1)
The function states that the output of the ALU will be equal to a signed minus 1. (All bits will be set).
    output answer 1111
    CN+4 = 1 (borrow)

EXAMPLE 5: (Select = 0100) (Function A Plus A$\overline{B}$)
The function states that the output of the ALU will be equal to word A ANDed to word B inverted and then added to word A.
    A = 1001
    B = 0110
    $\overline{B}$ = 1001
    A$\overline{B}$ = 1001
    A = 1001
    output answer 0010
    CN+4 = 0 (carry)

EXAMPLE 6: (Select = 0101) (Function (A + B) Plus A$\overline{B}$)
The function states that the output of the ALU will be equal to word A ORed to word B added to the result of word A ANDed to word B inverted.
    A = 1001
    B = 0110
    A+B = 1111
    A$\overline{B}$ = 1001
    output answer 1000
    CN+4 = 0 (carry)

EXAMPLE 7: (Select = 0110) (Function A Minus B Minus 1)
The function states that the output of the ALU will be equal to word A minus word B minus the number 1.
    A = 1001
    B = 0110
    A−B = 0011
    output answer 0010
    CN+4 = 0 (no borrow)

EXAMPLE 8: (Select = 0111) (Function A$\overline{B}$ Minus 1)

Digital Circuits and Devices

The function states that the output of the ALU will be equal to word A ANDed to word B inverted and 1 is subtracted from this result.
    A = 1001
    B = 0110
    $A\overline{B}$ = 1001
    output answer 1000
    CN+4 = 0 (no borrow)

EXAMPLE 9: (Select = 1000) (Function A Plus AB)

The function states that the output of the ALU will be equal to word A ANDed to word B. The results will be added to word A.
    A = 1001
    B = 0110
    AB = 0000
    output answer 1001
    CN+4 = 1 (no carry)

EXAMPLE 10: (Select = 1001) (Function A Plus B)

The function states that the output of the ALU will be equal to word A added to word B.
    A = 1001
    B = 0110
    output answer 1111
    CN+4=1 (no carry)

EXAMPLE 11: (Select = 1010) (Function $(A + \overline{B})$ Plus AB)

The function states that the output of the ALU will be equal to word A ORed to word B inverted. The results will then be added to the result of word A ANDed to word B.
    A = 1001
    B = 0110
    $A+\overline{B}$ = 1001
    AB = 0000
    output answer 1001
    CN+4 = 1 (no carry)

EXAMPLE 12: (Select = 1011) (Function AB Minus 1)

The function states that the output of the ALU will be equal to word A ANDed to word B and from this result 1 will be subtracted.
    A = 1001
    B = 0110
    AB = 0000
    output answer 1111
    CN+4 = 1 (borrow)

EXAMPLE 13: (Select = 1100) (Function A Plus A)
The function states that the output of the ALU will be equal to word A added to word A.
   A = 1001
   output answer 0010
   CN+4=0 (carry)

EXAMPLE 14: (Select = 1101) (Function (A + B) Plus A)
The function states that the output of the ALU will be equal to word A ORed to word B and the results added to word A.
   A = 1001
   B = 0110
   A+B = 1111
   output answer 1000
   CN+4=0 (carry)

EXAMPLE 15: (Select = 1110) (Function (A + $\overline{B}$) Plus A)
The function states that the output of the ALU will be equal to word A ORed to word B inverted and the result added to word A.
   A = 1001
   B = 0110
   A+$\overline{B}$ = 1001
   output answer 0010
   CN+4=0 (carry)

EXAMPLE 16: (Select = 1111) (Function A Minus 1)
The function states that the output of the ALU will be equal to word A minus 1.
   A = 1001
   output answer 1000
   CN+4=0 (no borrow)

As can be seen from these examples, the function of the ALU determines the output state of the ALU.

Topic Review   12-2

1. The _____ _____ _____ performs both mathematical and logic functions on data within a computer.

2. The simplest method of increasing the work size of an ALU is to use the _____ _____ operation.

3. The word size of an ALU can also be expanded by using a _____ _____ _____ .

Digital Circuits and Devices

4. The arithmetic logic unit can be considered the _____ section of a computer system.

5. The 74181 is a 4-bit TTL ALU that can perform sixteen _____ functions and sixteen _____ functions.

Answers:

1. arithmetic logic unit
2. ripple carry/borrow
3. look-ahead carry generator
4. calculating
5. logic, arithmetic

12-3 Introduction to the 6802 Microprocessor

An MPU (Microprocessor Unit) is a device usually in one integrated package that interprets instructions stored in memory in a sequential fashion and then manipulates that data to perform a specified task. The MPU is only the guidance part (or brain) of the computer system. It alone cannot perform very many tasks but when allowed to control other devices, the MPU can perform complex control operations. We will discuss Motorola's 6800 base 6802, which is a member of its 6800 family of MPUs. Of course, there are many different MPUs available today; however, the concepts are all basically the same as the 6802.

Before we begin talking about the computer system as a whole, we will learn about the MPU which makes up the controlling system of the computer itself. The 6802 is an 8-bit NMOS MPU that can interpret 72 different instructions. Its 7 different addressing modes allow each instruction to perform its task very efficiently depending on how it is used in the program. With its 16 unidirectional address lines, the 6802 can access up to 65,536 unmultiplexed memory locations. Finally the 6802 also contains 8 bidirectional data lines which allow data to be transferred inside the MPU and also throughout the computer system.

BLOCK DIAGRAM OF THE 6802 MPU

Shown in **Figure 12-3** is a block diagram of the 6802 MPU. The following provides a description of the function of each block within the 6802. By learning the function of each block in the MPU, the operation of the MPU as a whole should become easier to comprehend.

12
Introduction to Computers

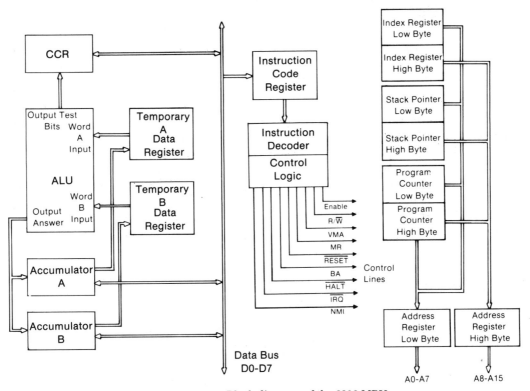

Figure 12-3. Block diagram of the 6802 MPU

ALU

The ALU is an 8-bit *Arithmetic Logic Unit*. Its purpose is to perform all arithmetic and logic manipulation for the MPU. The ALU in the 6802 has the ability to perform the following functions on two 8-bit words supplied by the data bus of the MPU.

Add
Subtract
AND
OR
ex-OR
Complement
Shift Right
Shift Left
Increment
Decrement

349

Digital Circuits and Devices

The ALU will supply an 8-bit answer which is transferred to one of the two accumulators of the MPU. There are also six output test bits used to determine special conditions relating to the value of the two 8-bit words or the results of the instruction performed. The six output test bits are supplied to the condition code register which will be studied next.

CCR

The *Condition Code Register* (CCR) is an 8-bit register used to hold the six test bits from the ALU. The two unused bits in the CRR (bit 6 and bit 7) are always set in the MPU. The MPU normally only reads this register for information on decision making. However, there are some instructions that allow this register to be written into. The following discussion explains what each bit represents in the CCR and how each bit changes its level. When the MPU is first powered up, all bits in the CCR are *low* except for the two MSBs which are not used and are always *high*. Shown in **Figure 12-4** is a diagram that shows the bit representation of the CCR.

Bit 7	Bit 6	Bit 5	Bit 4	Bit 3	Bit 2	Bit 1	Bit 0
1	1	H	I	N	Z	V	C

Figure 12-4. Diagram showing bit representation of the CCR

Bit 0: (Carry bit)
  This bit will set if a carry is generated by the MSB in an addition problem.
Bit 1: (Overflow bit)
  This bit will set if a 2's complement overflow is generated by a subtraction math problem. The 2's complement of a binary number is the binary number that results when 1 is added to the 1's complement. The 1's complement of a binary number is the number that results when each 0 is changed to a 1 and each 1 is changed to a 0. Instead of subtracting one number from another, you can add the 2's complement of the number to be subtracted and disregard the last carry.
Bit 2: (Zero bit)
  This bit will set if an operation results in an answer of zero.
Bit 3: (Negative bit)
  This bit will set if the MSB of the answer is set which represents a negative signed number. A signed number refers to when the MSB of a number is used to represent if the number is positive or negative. Normally, when the MSB is set, the number is negative; and when the MSB is reset, the number is positive.
Bit 4: (Interrupt mask bit)
  When this bit is set, interrupts will not be serviced by the computer system. This

bit will be set by the programmer or program.

Bit 5: (Half carry bit)

This bit will set if a carry is generated from the lower nibble to the high nibble during an arithmetic problem.

Bit 6 and Bit 7:

These two bits are not used by the MPU, and they are always set.

Temporary A & B Data Registers

The *temporary A & B data registers* are used to apply to the ALU so that the ALU function can be performed. The reason the two data registers are needed is the ALU has no storage capability. Both registers are 8-bits and can be written into by only the A and B accumulators. The outputs of the two registers will be immediately applied into the word A and word B inputs of the ALU.

AccA and AccB

These two accumulators (Accumulator A and Accumulator B) are 8-bit registers used to transfer data to and from the data bus into and out of the ALU. To perform manipulation on data, the data must first be transferred into one of the accumulators. After the manipulation has been performed, the ALU will store its answer into either AccA or AccB, depending on the instruction used.

Instruction Code Register

The *instruction code register* is an 8-bit register used to hold the binary code of the instruction that is being executed. The data bus acts as its only input and its output is applied to the instruction decoder circuit. The instruction code register is the first register to be loaded during the operation of a program.

Instruction Decoder

The *instruction decoder* takes the 8-bit instruction code from the instruction code register and produces a logic control for the control logic circuitry.

Control Logic Circuitry

The *control logic circuitry* controls the sequence of operation for each block in the MPU IC. Its function includes the transfer of data throughout the MPU, and control of the external control lines for the MPU which allows the MPU to control the entire computer system.

Digital Circuits and Devices

ID

The *InDex register* is a 16-bit register. It can be used to modify data or memory locations depending on how the programmer uses the register.

SP

The *Stack Pointer* is a 16-bit register that will point to an address location in RAM that is used to store the internal environment of the MPU when an interrupt occurs. The interrupt will cause the data in the InDex register, Program Counter, AccA, AccB, and the CCR to be stored in RAM until the interrupt routine is complete. Then, the MPU will reload all the data from RAM back into the proper registers. The reason for the storage of the original data is to allow the MPU to restart its main program from where it was interrupted. This does not allow any changes made in the internal environment during the interrupt routine to affect the main program.

Address Register

The *address register* is a 16-bit register that contains the address of the memory location that is being accessed at any given time. The control logic circuitry will load data from the data bus into the address register which will be used as the address for the address bus.

6802 MPU Control Lines

The control logic circuitry controls the levels of the control lines which acts as outputs. The levels on the output control lines will be determined by the instruction that is being executed. Some of the control lines act as inputs. In this case, an external device will control the level on the control line, and the control logic circuitry inside the MPU will determine the action that is to be taken. There are 9 control lines for the 6802. The following outlines the function of each line.

Enable

The Enable (clock line) output is the second phase of a two phase clock. This line allows external devices in the computer system to become enabled during the second phase of the clock. During the first phase of the clock, only the MPU is enabled. The clock uses a 3.579545 *MHz* crystal to generate a 894.8 *kHz* two phase clock.

R/$\overline{W}$

The Read/Write output line is used to control whether devices will be read from or written into by the MPU. When this line is *high* the MPU will read data from the selected memory lcoation. When this line is *low* the MPU will write data into the selected memory lcoation.

VMA

The Valid Memory Address output line is a signal that tells a device there is a valid memory address on the address bus when this line is *high*.

MR

The Memory Ready output line is used to control slow memory devices. Therefore, invalid data will not be on the data bus when the device or the MPU is ready to use the data.

RESET

The RESET input line is a hardware line connected to a key on the keyboard that is used to manually reset the MPU and all devices in the system. This line will reset the system when brought *low*. When *high* the system will operate in a normal manner.

BA

The Bus Available output line allows external devices to use the address bus when the MPU is not using the bus. As a result, the external device can place an address on the address bus. This line can be used by a device like a Direct Memory Access Controller which allows the MPU not to waste time controlling devices that do not need MPU control. This will make the operation of the MPU more efficient.

HALT

The HALT input line is controlled by an external device that will cause all operations of the MPU to stop until the line goes *high* again. This line allows external devices to control the MPU when the external device must maintain control of the system.

IRQ

The Interrupt ReQuest input line is used by an external device to cause the MPU to start its interrupt sequence. If the interrupt mask bit in the CCR is set, this input

Digital Circuits and Devices

will not cause the interrupt sequence to occur.

$\overline{\text{NMI}}$

The Non-Maskable Interrupt request input line is used by an external device to cause the MPU to start its interrupt sequence. The condition of the interrupt mask bit in the CCR or any other signal will have no effect in stopping the interrupt sequence when NMI line is activated.

PROGRAM OPERATION EXAMPLE

Now that the operation of each block inside the MPU is known, it's time to see the sequence of each block inside the MPU during the operation of a simple program. The following is a program that will load AccA with the data from memory location $7F (the $ sign indicates the number is a hexadecimal value). Then add the data found in AccA to the data found in memory location $7E. The answer of the addition will be stored in AccA. At this point, the program will stop.

PROGRAM FOR USER:	PROGRAM FOR COMPUTER:	
LDAA from MLoc 7F	$96	$7F
ADDA from MLoc 7E	$9B	$7E
SWI	$3F	

In the two examples of the same program above, the example of the user program uses English type statements to determine what each instruction does. However, the computer can only understand binary data. Therefore, the numbers for the computer program example are entered into the computer in hexadecimal. The computer will see each number as four binary bits, and it will use the binary data to perform the instruction.

LDAA

The LDAA instruction in our example will load Accumulator A with the data found in memory location 7F.

ADDA

The ADDA instruction in our example will ADD the data found in accumulator A to the data found in memory location 7E.

SWI

The SoftWare Interrupt instruction will end the program, and it will display the data found in the Program Counter to indicate to the user that the program is complete.

To run the program, the user must enter the address of the first byte of the program, and then press the [GO] key. The control logic will load the address into the address register which will cause the program counter to increment by one. The control logic will then cause the data in the selected memory location to be applied to the data bus. Next, the control logic will cause the data on the data bus to be transferred into the instruction code register. The data now in the instruction code register is applied to the instruction decoder. This decoder will apply the proper signals to the control logic to execute the following task. The data in the program counter will be loaded into the address register to become the new address. This will cause the program counter to increment by one. Next, the data at the selected address will be stored in the lower byte of the address register. With the new address in the address register, the data at the selected memory location will be loaded into accumulator A. The control logic will now transfer the data in the program counter into the address register causing the program counter to increment by one. Data at the new address will be transferred into the instruction register. This binary data will be decoded by the instruction decoder, and it will cause the control logic to perform the following tasks. The data in the program counter will be loaded into the address register, and the program counter will be incremented by one. Data at the selected memory location will be applied to the data bus. This data will then be transferred into the word B temporary data register of the ALU by the AccB (note the data will not be stored in AccB). After the addition is performed in the ALU, the answer will be transferred into the AccA. The answer of the addition will remain in AccA until an instruction is given to cause the data to change locations. The control logic will now transfer the data in the program counter into the address register and increment the program counter by one. Data at the selected memory location will be transferred into the instruction code register. This will be decoded by the instruction decoder, and it will cause the control logic to perform the following task. The last instruction will cause the data in the program counter to be applied to seven segment displays, and it will halt any further operation of the MPU. This ends the example program given. Normally the answer that is stored in AccA will be used to perform some other function when the program is larger.

In the preceding explanation of the operation of the example program, it should be noted that the control logic knows when the instruction is complete. Also at the completion of the instruction, the control logic will load the next byte of data into the instruction code register. The instruction code register then uses this data as the next new instruction. This process will continue until the instruction code register receives

Digital Circuits and Devices

the code to stop the program.

Topic Review 12-3

1. An _____ is a digital device that interprets various instructions stored in memory in a sequential manner to perform a specified task.

2. The 6802 is an _____ NMOS MPU that can interpret _____ different instructions and provide _____ different addressing modes.

3. The 6802 MPU contains two _____ which are used to transfer data to and from the data bus and into and out of the ALU within the MPU.

4. The _____ _____ circuitry controls the sequence of operation for each block within the MPU IC.

5. The _____ _____ is a 16-bit register that contains the address of the memory location that is being accessed at any given time.

Answers:

1. MPU
2. 8-bit, 72, 7
3. accumulators
4. control logic
5. address register

12-4 6802 Computer System

As stated earlier in the chapter, the MPU can only manipulate data that is applied to it. In addition, the MPU itself has very little memory capability (128 bytes). Therefore, to cause the MPU to operate as a *computer system*, there will be a need for *support hardware*. The support hardware usually includes additional memory, some type of I/O (Input/Output) device, and some type of program stored in ROM. These devices will tell the MPU how to get data and how to send data to the I/O device. The program that will cause the MPU to act like a computer is called the *monitor program* (in the D5 kit it is identified as D5BUG). The monitor program is required for a computer system to operate. Shown in **Figure 12-5** is the basic block diagram of the D5 Kit 8-bit microcomputer system.

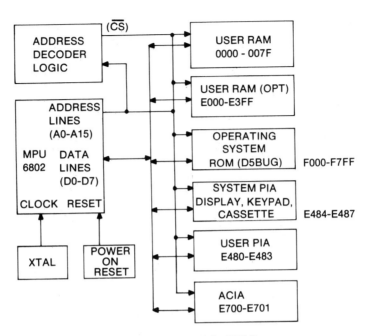

Figure 12-5. Block diagram of the D5 kit

MPU

The 6802 is an 8-bit MPU that can manipulate the data applied to its *data bus*. The 16-bit address bus and the 9 control lines controlled by the MPU will determine the data that will appear on the 8-bit data bus. The 6802 has 72 different instructions, and it provides a wide range of tasks available for programming purposes.

XTAL

The crystal circuit provides the necessary input frequency to the 6802 which will produce the two phase clock signal needed to operate the computer system.

POWER ON RESET

The power on reset circuit will cause a reset to occur in all devices of the computer system when power is first applied to the circuit. This condition will give the computer system a starting point for reference.

ADDRESS DECODER LOGIC

The address decoder logic circuit uses some of the address lines to access blocks of memory in the computer system. The outputs of the address decoder logic are connected to the chip select pins for each block of memory. This allows many different size memory devices to be connected together to function in the 8-bit computer system.

USER RAM

There are two blocks of user RAM in the D5 kit microcomputer system. Block 1 of user RAM uses memory locations 0000-007F in hex, and it is mandatory for the minimum operating system. The second block of user memory is optional. This block occupies memory locations E000-E3FF in hex. The term user RAM simply applies to RAM that the user can use for a program.

OPERATING SYSTEM

The operating system (monitor program, identified as D5bug in the D5 kit) is a program that is stored in ROM. Therefore, when power is first applied to the circuit, the program will automatically start executing. This program tells the MPU how to act as a computer. It occupies memory locations F000-F7FF in hex.

PIA

There are two PIAs in the D5 kit. One is called the system PIA (memory locations E484-E487), and the other is called the user PIA (memory locations E480-E483). The system PIA is used to control data to and from the computer displays, keypad, and cassette interface, so the MPU can communicate with the outside world in a parallel fashion. The system PIA's operation is controlled automatically by the MPU under the guidance of the monitor program. The user PIA is used by the user under control of the program being executed. Note this PIA does not have to be used.

ACIA

The ACIA is an optional I/O device that allows the computer system to communicate with the external world through asynchronous devices and is controlled by the users program. The ACIA is controlled by the data in memory locations E700-E701.

MEMORY MAP

From the block functions of the computer system it should be noted that each device occupies a certain location in memory. A memory map allows the user to easily locate the address location of any device in the computer system. **Figure 12-6** illustrates the memory map for the D5 kit.

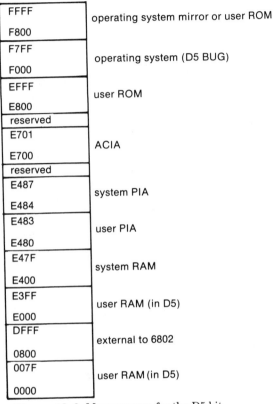

Figure 12-6. Memory map for the D5 kit

Topic Review 12-4

1. In order for an _____ to operate as a computer system, there is a need for support hardware.

2. The support hardware usually includes additional _____ , some type of _____ device, and some type of program stored in _____ .

3. The _____ _____ logic circuit uses a few address lines to access blocks of memory.

4. _____ _____ refers to RAM that the user can store a program.

Digital Circuits and Devices

5. The _____ _____ will tell the MPU how to get data and how to send data to an I/O device.

Answers:

1. MPU
2. memory, Input/Output, ROM
3. address decoder
4. User RAM
5. support memory

12-5 Summary Points

1. The ALU (Arithmetic Logic Unit) is the part inside a computer IC, that performs all mathematical and logic functions.

2. When expanding the word size of the 74181 ALU, the carry function must be transferred between the ALUs. A serial carry is performed when the CN + 4 output is applied to the CN input of the next ALU. This method becomes slower as the word size increases. A parallel carry is performed when the propogate and generate pins of the ALUs are used in conjunction with the 74182 look ahead carry generator. This method of producing a carry provides a speed of operation that is independent of word size.

3. Most computer terminology is the same even though different computers are used.

4. An MPU is a device that interprets instructions stored in memory in a sequential order to perform a specified task. Due to its limitations, additional hardware is needed to make it operate as a complete computer system.

5. The address bus of a computer system acts as an output for the MPU IC, and it acts as an input to all other devices in the computer system.

6. The data bus of a computer system can act as an input or an output to any device in the computer system.

7. The Condition Code Register (sometimes called the Status Register) contains test bits of the ALU function that are used in making decisions in the computer program.

12
Introduction to Computers

8. The monitor program is necessary in all computer systems in order to have a computer operate properly.

9. All computers must have some sort of peripheral interface adapter (PIA) that allows the computer to communicate with the outside world in a parallel fashion.

10. Most computers have some sort of asynchronous communication interface adapter (ACIA) that is used to allow the computer to communicate with the outside world through serial devices.

12-6 Chapter Progress Evaluation

1. Can the 74181 ALU perform multiplication and division functions?

2. When the 74181 is used with the 74182 look ahead carry generator, what type of carry is the ALUs carry bit?

3. When word A has the same value as word B in the 74181 ALU which output will always go *high*?

4. What is the name of the digital device that converts parallel data in serial data for the outside world to use?

5. A signal to the MPU that causes the MPU to halt current program execution, save the internal environment, and execute a program to service the signal, is called?

6. The control line of the 6802 that controls whether data will be stored or retrieved from a memory device, is called?

7. The monitor program is stored in which type of memory device?

8. The PIA in the 6802 based computer system allows the computer to communicate with the outside world in what type of data format?

9. The output lines of the 6802 that selects which memory location will be accessed are called?

10. The register inside the 6802 that holds the code of instruction being executed, is called what?

Index

Digital Circuits and Devices

ACC (accumulator) 337, 351, 352
access time 293
accumulator: *see* ACC
ACIA (asynchronous communication interface adapter) 337, 358
ADCs (analog-to-digital converters) 269, 279-290
   counter 286-287
   integration 283-285
   parallel 288-289
   servo 286
A/D (analog-to-digital) conversion 279-290
addend 158
adders 156-159
   cascading of 159
addition, binary 155-159
address 293
   bus 337-338
   register 352
algebra, Boolean 7-8
ALU (arithmetic logic unit) 337, 338-347, 349-350, 355
American Standard Code of Information Interchange: *see* ASCII
analog
   -to-digital conversion: *see* A/D conversion
   -to-digital converters: *see* ADCs
   and digital systems contrasted 8-9
   signals 1-3
   systems 3
ANDed variables: *see* minterms
AND gate 6, 34-42, 130-131
   truth table 34, 35ff
AND identities, Boolean 130-131
arithmetic operations of digital circuitry 155-160
ASCII (American Standard Code of Information Interchange) code 29-30, 175, 181-182
associative and commutative laws 137-138, 142
associative law 136
async: *see* asynchronous
asynchronous
   communication interface adapter: *see* ACIA
   counter 227-236. *See also* ripple counter and synchronous counter contrasted 249
augend 158
avalanche current 104

BA (Bus Available) 353

base 8 numbering system: *see* octal numbering system
base 16 numbering system: *see* hexadecimal (hex) numbering system
base 10 numbering system: *see* decimal numbering system
base 2 numbering system 11
BCD codes 26-28
bilateral switches (digital) 186
binary
   addition 155-159
   -coded decimal (BCD) numbers. *See also* BCD
   counting 12
   to decimal conversion 14
   and decimal and hexidecimal numbering systems compared 17
   and decimal systems compared 12-16
   numbering system 11-16
   and octal numbering systems compared 23
   subtraction 156, 159-160
   weights 12-13
bipolar transistor ROM 296, 302-304
BI/RBO (Blanking In/Ripple Blank Out) 180
bit position weights: *see* binary weights
bits 11
Blanking In/Ripple Blank Out: *see* BI/RBO
Boole, George 8, 129
Boolean
   algebra 7-8, 129ff
   AND identities 130-131
   expression of gate function 36, 37
   expressions 75-90
   inverter (NOT) identities 134-135
   laws 136-138
   NOT identities 134-135
   OR identities 132-133
   theorems 139-142
bubble memory: *see* MBM
burn-in (ROM) 297
burst refresh mode 326
bus, address 337-338
bus, data 299, 338
Bus Available: *see* BA
bytes 19, 293

carry generator, 74182 look-ahead 340
cascading

Digital Circuits and Devices

of adders 159
of counters 259
of subtractors 160
cause/stop method of non-standard synchronous counter design 251-257
CCD (charge coupled device) 329
CCR (condition code register) 339, 350-351, 352
charge coupled device: *see* CCD
check bit: *see* parity
Chip Select: *see* CS
circuitry, minimization of 146-147
clocked RS latches 199-203
clock rate: *see* propagation delay
CML (current-mode logic) 92-93, 114, 121
CMOS (complementary MOS) 118-121
  memory 323
code(s). *See also* coding systems
  ASCII 29-30, 175, 181-182
  BCD 26-28
  EBCDIC 30-31
  8421 (BCD) 27
  excess-3 (BCD) 27-28
  Gray 29
coding systems 29-31. *See also* codes
communative law 136-137
comparators, magnitude 186-187
complementary MOS: *see* CMOS
computers 337-361
computer terminology 337-339
condition code register: *see* CCR
control logic (MPU) 338, 351
conversions: *see under* system name, as *binary to decimal conversion*
converter, flash: *see* parallel ADC
converter, modified parallel 289
counter(s) 227-267. *See also* down counters, up counters, up/down counters
  ADCs 286-287
  cascading of 259
  mod 2 228-229
  mod 3 250-254
  mod 4 229-233, 238-249
  mod 8 233-236
  mod 16 246-247
  non-standard synchronous 250-260
    design of 250-257
  pseudo random 260-266
  shift register 222-224
  standard synchronous 237-249
CS (Chip Select) 294, 325-326
  circuit 316-317
current-mode logic: *see* CML
cycle time 294

D/A (digital-to-analog) conversion 269-279
DAC(s) (digital-to-analog converters) 269-279, 282
  resolution 272

testing 276-279
data
  bus 299, 338, 357
  flip-flops 214-215
  latches 204-206
    quad (7477) 205-206
  registers 351
  selector: *see* multiplexer
decimal
  to binary conversion 14-16
  and binary and hexadecimal numbering systems compared 17
  and binary numbering systems compared 12-16
  to hexadecimal conversion 21
  numbering system 12-16
  weights 12-13
decoders 167-169. *See also* decoding; demultiplexers
  instruction 338, 351
decoder/driver, built-in 177
decoder/driver, display *see* display decoder/driver
decoding 26, 166-169. *See also* decoders
delay, propagation 93-95
"DeMorganization" 141
DeMorgan's law 138, 141-142
demultiplexers 164-166
  TTL 166
depletion-mode MOS devices 117
detector: *see* decoders
digital
  -to-analog conversion: *see* D/A conversion
  -to-analog converters: *see* DACs
  and analog systems compared 8-9
  concepts 1-10
  displays 171-183
  gates 6-7, 33-74
  integration 155-190
  signals 4
  systems 4-5
  terminology 5-7
diode
  matrix ROM: *see* diode ROM
  ROM 297-302
  Schottky (barrier) 108
  -transistor logic: *see* DTL
  zener 104
Disk Operating System: *see* DOS
display(s)
  decoder/driver 178-183
  digital 171-183
  formats 173-177
distributed refresh mode 326
distributive law 138, 141-142
divide-by-two circuit; *see* mod 2 counter
DMOS (doping MOS) 118
doping MOS: *see* DMOS
DOS (Disk Operating System) 334

Index

dot matrix decoder/driver 181-183
dot matrix digital display 175-177
down counter
   mod 4 230-232, 239-240
   mod 8 233-234, 244-245
   mod 16 246-247
DTL (diode-transistor logic) 92, 94, 103-104, 121
dual-ramp ADC: *see* integration ADC
dual-slope ADC: *see* integration ADC
dynamic memory 294
dynamic RAM 323-327

EAPROM (Electrically Alterable Programmable Read Only Memory): *see* EEPROM
EBCDIC (extended binary-coded decimal interchange code) 30-31
ECL (emitter-coupled logic) 92-96, 98, 114-115, 121, 122
   memory 323
EEPROM (Electrically Erasable Programmable ROM) 297
8421 BCD code 27
Electrically Erasable Programmable ROM: *see* EEPROM
Electrically Alterable Read Only Memory (EAPROM): *see* EEPROM
emitter-coupled logic: *see* ECL
Enable 352
encoders 169-170. *See also* encoding
encoding 26, 166-167, 169-170. *See also* encoders
enhancement-mode MOS devices 117
environment, internal (computer) 338
epitaxial garnet 331
EPROM (Erasable Programmable Read Only Memory) 294, 297, 311-313
Erasable Programmable Read Only Memory: *see* EPROM
excess-3 BCD code 27-28
exclusive-NOR (ex-NOR) gates 64-66
exclusive-OR (ex-OR) gates 61-64
ex-NOR: *see* exclusive-NOR
ex-OR: *see* exclusive-OR
expandable gate function 184-185
extended binary-coded decimal interchange code: *see* EBCDIC

fan-in 97-98, 102
fan-out 97-98, 102, 109-110, 121
Fetch operation: *see* Read operation
flash converter: *see* parallel ADC
flip-flops 33, 206-224, *passim*
   data 214-215
   JK 228
   and latch compared 207
   master slave 206-214
floppy disk storage 334-335

forced reset method of non-standard synchronous counter design 250-251
formats, display 173-177
full adders 157-159
full subtractors 159-160

gadolinium gallium garnet: *see* GGG
gas discharge tubes 171, 173-174
gate(s)
   AND 34-42, 130-131
   circuits, derivation of 80-88
   digital 6-7, 33-74
   exclusive-NOR (ex-NOR) 64-66
   exclusive-OR (ex-OR) 61-64
   inversion 66-71
   NAND 51-55
   negative-logic 122-125
   NOR 56-61
   NOT: *see* inverter
   OR 42-49, 132-133
generator/checker, parity 187-189
GGG (gadolinium gallium garnet) 331
Gray code 29

half adders 156-157
half subtractors 159
HALT line 353
HCMOS (high performance CMOS) 119-120
hex: *see* hexadecimal
hexacecimal
   to binary conversion 18-20
   and binary and decimal numbering systems compared 17
   counting 17-18
   to decimal conversion 20
   numbering system 17-24
   to octal conversion 23-25
hidden refresh 326
high output method (of deriving gate circuits from truth tables) 86-87
high performance CMOS: *see* HCMOS
high performance MOS: *see* HMOS
high-speed TTL 92, 106-107, 110
high-threshold logic: *see* HTL
HMOS (high performance MOS) 118
HTL (high-threshold logic) 92, 96, 104

ID (index register) 338, 352
IIL: *see* I^2L
immunity, noise 95-96
index register: *see* ID
Input/Output: *see* I/O
instruction code register 351
instruction decoder 338, 351
instruction register 338
integrated-injection logic: *see* I^2L
integration ADC 283-285

365

integration, digital 155-190
interfacing logic families 120-121
internal environment (computer) 338
internal operational truth table and logic truth table contrasted 34-35
Interrupt 338
　Request: see IRQ
inversion, gate 66-71
inverter(s) 7, 49-51
　identities, Boolean 134-135
　NAND 55-56
　NOR 60-61
I/O (Input/Output) device 356
IRQ (Interrupt Request) 353-354
I²L (integrated injection logic) 92-94, 98, 115-116, 121

JK flip-flops 228
　master slave 210-214
Johnson counter 223-224
Josephson Junction memory 329-330

Karnaugh
　loops 150-151
　map(ping) 143-152, 261-262, 264-265
　　for circuit minimization 146-147
　　for plotting multiple variables 148-149
　　rolling of 149-150
　　and truth table compared 145-146
K-map: see Karnaugh map(ping)

ladder, R2R
large loop MBM system 332-333
large-scale integration: see LSI
latch(es) 191-206
　data 204-206
　and flip-flop compared 207
　RS 191-206
law(s)
　associative 136
　　and commutative 137-138, 142
　Boolean 136-138
　commutative 136-137
　DeMorgan's 138, 141-142
　distributive 138, 141-142
LCD (liquid crystal display) 172-174
LEDs (light emitting diodes) 171-177
light emitting diodes: see LEDs
light generating digital displays 171-172
liquid crystal display: see LCD
loading (factor) 97-98
logic
　circuitry, minimization of 146-147
　development 75-90
　families 91-128
　　characteristics of 121

interfacing 120-121
negative 122-125
positive and negative compared 6
simplification 129-154
translators 98
truth table and internal operational truth table contrasted 34-35
long loop CCD 329
look-ahead carry generator (74182) 340
looping (Karnaugh) 149-151
loop value 144
low output method (or deriving gate circuits from truth tables) 87-88
low-power Schottky TTL 92, 94, 97, 109-110. See also low-power TTL; Schottky TTL
low-power TTL 92, 94, 107-108, 110. See also low-power Schottky TTL
low threshold PMOS 118
LSB code 181
LSI (large-scale integration) 116

magnetic
　bubble memory: see MBM
　core memory 330
　memory storage 330-335
　tape storage 334
magnitude comparators 186-187
major/minor loop MBM system 332-333
map(ping), Karnaugh: see Karnaugh map(ping)
map, memory 358-359
mask, ROM 296-297
master slave flip-flops 206-214
MBM (magnetic bubble memory) 331-334
memory 293-336
　address register 338
　CMOS 323
　ECL 323
　expansion 327-329
　map 358-359
　non-volatile 295, 296
　static 295
　volatile 295
Memory Ready: see MR
metal-oxide-semiconductor: see MOS
micro: see MPU
microprocessor: see MPU
minimizing logic circuitry 146-147
minterms 143, 149
mod 227
mod counter, non-standard: see non-standard synchronous counter
mod 8 counters 233-236
　down counters 233-234, 244-245
　up counters 233-234, 242-244
　up/down counters 233, 235, 245-246
mod 5 up counter 254-257

mod 4 counters 229-233
  down counters 230-232, 239-240
  synchronous 238-249
  up counters 229-230, 238-239
  up/down counters 232-233, 240-241
modified parallel converter 289
modified short loop CCD 329
mod 9 up counter 258
mod 7 up counter 257-258
mod 6 up counter 257-259
mod 16 counters 246-247
  down counters 246-247
  up counters 246-247
  up/down counters 246-249
mod 10 up counter 258-259
mod 3 counter 250-254
  up counter 253-254, 259
mod 2 counter 228-229
modulo division 21. *See also* decimal to binary conversion
modulus, counter: *see* mod
Monitor program 338, 356-359
monotonicity test (DAC) 276-279
MOS (metal-oxide-semiconductor) 91ff, 117-120
  FET 296
  ROM 304-308
  RAM 323-327
    static 321-323
  ROM: *see* MOS FET ROM
  use of 121
MPU (microprocessor unit) 338-339, 348-359
  6802 348-359
MR (Memory Read) 353
MSB code 181
multiple
  dot matrix
    decoder/driver 183
    digital display 177
  7-segment digital display 174-175
  variables, plotting with Karnaugh map 148-149
multiplexed decoder/driver 180-181
multiplexer/decoder/driver, 7-segment 174-175. *See also* display decoder/driver
multiplexers 161-164
  TTL 164

NAND
  gates 51-56
    as inverters 55-56
  inverters 55-56
  latch
    clocked RS 201-203
    RS 195-198
N-channel MOS: *see* NMOS
negative
  AND gates 122-123, 125

  logic 6, 122-125
    gates 122-125
    -OR gates 122-124
nematic gell 172-173
nibbles 19, 294
NMI (Non-Maskable Interrupt) 354
NMOS (*N*-channel MOS) 117-118, 120, 121
noise (as a factor in digital circuitry) 95-96
noise immunity 95-96
non-light generating digital displays 172-173
Non-Maskable Interrupt: *see* NMI
nonsaturated bipolar logic 91ff, 114-116
non-standard synchronous counters 250-260
non-volatile memory 295, 296
NOR
  gates 56-61
  inverter 60-61
  latch
    clocked RS 199-201
    RS 192-195
NOT
  AND: *see* NAND
  gate: *see* inverter
  identities, Boolean 134-135
  OR: *see* NOR
numbering systems 11-32

octal
  to binary conversion 22-23
  and binary numbering systems compared 23
  counting 22
  to decimal conversion 24-25
  to hexadecimal conversion 23-25
  numbering system 22-26
1-of-8 decoders 167-169
open collector TTL 111-112
OR
  gates 6, 42-49, 132-133
  identities, Boolean 132-133
orthogonal coils 332

parallel ADC 288-289
parallel adders 156, 159
parameters of logic families 92-98
parity 187-189
  generator/checker 187-189
PC (program counter) 339, 352
P-channel MOS: *see* PMOS
peripheral interface adapter: *see* PIA
PIA (peripheral interface adapter) 339, 358
  user 358
pinout diagrams 101
plotting multiple variables with Karnaugh map 148-149
PMOS (*P*-channel MOS) 117-118, 120, 122
positive logic 6

367

Digital Circuits and Devices

program counter: *see* PC
Programmable Read Only Memory: *see* PROM
program-symbol notation 19-20
progressive subtraction: *see* decimal to binary conversion
PROM (Programmable Read Only Memory) 295, 297, 309-311
  burner 310
propagation delay 93-95
pseudo random counters 260-266
pull-up resistor 111
pull-up transistor 106, 107

quad data latch 205-206
quad-slope principle 285
quantization error, DAC 274

RAM (Random Access Memory) 266, 295, 315-327, 339, 352
  dynamic 323-327
  user 358
ramp-type ADC: *see* integration ADC
random access 296
Random Access Memory: *see* RAM
RAS only refresh 326
RBI (Ripple Blanking In) 179-180
Read Mostly Memory: *see* RMM
Read Only Memory: *see* ROM
Read operation 295
Read Write Memory: *see* RWM
READ/WRITE time 294
reduction, variable 139-142
refresh operation 294, 295, 323, 325-327, 329
reglow 310-311
RESET line 353
reset/set latches: *see* RS latches
resistor-transistor logic: *see* RTL
resolution, DAC 272
ring counter 223
Ripple Blanking In: *see* RBI
ripple carry 339-340
ripple counter 235-236
RMM (Read Mostly Memory) 295, 309-314
rolling (a Karnaugh map) 149-150
ROM (Read Only Memory) 266, 295, 296-309, 338, 356, 358
  bipolar transistor 302-304
  diode 297-302
  ICs 307-308
  MOS FET 304-308
RS
  latches 191-206
    clocked 199-203
  master slave flip-flops 206-209
  NAND latch 195-198
    clocked 201-203

NOR latch 192-195
  clocked 199-201
RTL (resistor-transistor logic) 92, 94, 102-103, 121
R2R ladder 275-276
RWM (Read Write Memory): *see* RAM

SA (successive-approximation) ADC 281-283j
SAM (Sequential Access Memory) 295, 334. *See also* CCD
sample-and-hold circuits 289-290
sapphire (as insulator in SOSCMOS technology) 120
saturated bipolar logic 91ff, 102-113
Schmitt trigger 185-186
Schottky (barrier) diode 108-109
Schottky TTL 92, 94, 108-110. *See also* low-power Schottky TTL
select lines, multiplexer 161-163, 164, 166
sense inversion 68-69
sequence counter: *see* ring counter
sequential access memory: *see* SAM
sequential counters: *see* counters
serial adder 156
servo ADCs 287
7-segment digital displays 173-175
7-segment multiplexer/decoder/driver 174-175
7448 decoder/driver 179-180
7447 decoder/driver 178-180
74181 4-bit MSI TTL ALU 339-347
74182 look-ahead carry generator 340
7477 quad data latch 205-206
shift register 215-224
  counters 222-224
shift right shift registers 216-219
short loop CCD 329
sign inversion 67-68
sign & sense inversion 70-71
simplification, logic 129-154
single-ramp converter: *see* integration ADC
single 7-segment decoder/driver 178-180
single 7-segment digital display 173-174
SISPO shift right shift register 216-219
6802 control lines 352-354
6802 MPU 348-359
SOSCMOS 119-120
SP (stack pointer) 339, 352
specs (specfications), logic family 98-102
SPISPO shift right register 219-221
split-octal notation 25-26
spreading factor 97
stack pointer: *see* SP
standard
  asynchronous counters: *see* asynchronous counters
  synchronous counters 237-249
  TTL 93, 105-106, 109, 110

static
  bipolar TTL RAM 315-321
  memory 295
  MOS RAM 321-323
  test (DAC) 276
  TTL RAM Read operation 318-320
  TTL RAM Write operation 320-321
status register: *see* CCR
Store operation: *see* Write operation
strobe lines, multiplexer 161-163, 164, 166
subroutine 339
subscript notation 19-20
subtraction, binary 156, 159-160
subtraction, progressive: *see* decimal to binary conversion
subtractor(s) 159-160
  cascading of 160
successive-approximation ADC: *see* SA ADC
switching time (rate): *see* propagation delay
sync: *see* synchronous
synchronous counters
  and asynchronous counters contrasted 249
  non-standard 250-260
    design of 250-257
  standard 237-249

temperature (as factor in digital circuitry) 96-98
terminology, digital 5-7
theorems, Boolean 139-142
tied logic: *see* wired logic
timing diagrams 38-39
toggle rate: *see* propagation delay
totem-pole output 106, 107
transistor-transistor logic: *see* TTL
translators, logic 98
tri-state TTL 112-113
truth tables 7, 34ff
  AND gate 34ff
  complex 83-88
  internal operational and logic contrasted 34-35
  and Karnaugh maps compared 145-146
  negative-logic 123-125
TTL (transistor-transistor logic) 92-98, 105-113, 120-121
  demultiplexers 166
  IC counters 260
TTL (*cont.*)
  multiplexers 164
  open collector: *see* open collector TTL
  74193 4-bit up/down counter 247-249
  7493 binary 4-bit ripple counter 235-236
  standard: *see* standard TTL
  tri-state: *see* tri-state TTL

up counter
  mod 3 253-254, 259
  mod 4 229-230, 238-239
  mod 5 254-257
  mod 6 257-259
  mod 7 257-258
  mod 8 233-234, 242-244
  mod 9 258
  mod 10 258-259
  mod 16 246-247
up/down counter
  mod 4 232-233, 240-241
  mod 8 233, 235, 245-246
  mod 16 246-247
user PIA 358
user RAM 358

Valid Memory Address: *see* VMA
variable reduction 139-142
variables, ANDed: *see* minterms
variables, multiple - plotting with Karnaugh map 148-149
V/F (voltage-to-frequency) ADCs 286
VMA (Valid Memory Address) 353
VMOS 118
volatile memory 295
voltage-to-frequency ADCs: *see* V/F ADCs

waveform synthesizer, DAC as a component of 271
WE (Write Enable) 296, 325-326
  circuit 316
weighted-resistor network: *see* WRN
wired logic 111-112
Write Enable: *see* WE
Write operation 296
write/recirculate shift register 221-222
WRN (weighted-resistor network) 271-274
  formula 273

XTAL 357
XY plotter, DAC-controlled 270

yes/no logic 129

zener diode 104
zener voltage 104